MW00333165

Luminos is the open access monograph publishing program from
UC Press. Luminos provides a framework for preserving and
reinvigorating monograph publishing for the future and increases
the reach and visibility of important scholarly work. Titles
published in the UC Press Luminos model are published with
the same high standards for selection, peer review, production,
and marketing as those in our traditional program.
www.luminosoa.org

The publisher gratefully acknowledges the generous support of the Ahmanson Foundation Humanities Endowment Fund of the University of California Press Foundation.

Hokum!

Hokum!

The Early Sound Slapstick Short and Depression-Era Mass Culture

———

Rob King

UNIVERSITY OF CALIFORNIA PRESS

University of California Press, one of the most distinguished university presses in the United States, enriches lives around the world by advancing scholarship in the humanities, social sciences, and natural sciences. Its activities are supported by the UC Press Foundation and by philanthropic contributions from individuals and institutions. For more information, visit www.ucpress.edu.

University of California Press
Oakland, California

Suggested citation: King, Rob. *Hokum! The Early Sound Slapstick Short and Depression-Era Mass Culture*. Oakland: University of California Press, 2017. doi: https://doi.org/10.1525/luminos.28

Library of Congress Cataloging-in-Publication Data

King, Rob, 1975– author.
   Hokum! : the early sound slapstick short and Depression-era mass culture / Rob King.
   Oakland, California : University of California Press, [2017] | Includes bibliographical references and index.
      LCCN 2016049875 (print) | LCCN 2016052138 (ebook) |
   ISBN 9780520288119 (pbk. : alk. paper) | ISBN 9780520963160 (Epub)
      LCSH: Comedy films—History and criticism. |
   Comedy films—United States—20th century.
   LCC PN1995.9.C55 K547 2017 (print) | LCC PN1995.9.C55 (ebook) |
   DDC 791.43/6170973—dc23
   LC record available at https://lccn.loc.gov/2016049875

25   24   23   22   21   20   19   18   17
10   9   8   7   6   5   4   3   2   1

To my parents

CONTENTS

LIST OF ILLUSTRATIONS AND AUDIOVISUAL MEDIA

FIGURES

VIDEO CLIPS

AUDIO CLIPS

ACKNOWLEDGMENTS

I have sometimes had the feeling that I was working on this book in secret, the private hobby of a decade's worth of summers. But I'll chalk that delusion up to the solitude of writing. I won't say that it took a village exactly, but it did take something like a campus or two.

The book has its genesis in 2006, in the chance confluence between my purchase of a DVD box set of Buster Keaton's Columbia shorts and Charles Wolfe's invitation to contribute an essay on Keaton to a special issue of *New Review of Film and Television Studies* (vol. 5, no. 3). I don't think he knows it, but it was Chuck who set me on the path that has eventually led here. Along the way, I have been supported by my colleagues and supervisors at the University of Toronto (where this project was started) and Columbia University (where it ended). I am particularly grateful to Elspeth Brown, Charlie Keil, and Janet Paterson at Toronto, and Nico Baumbach, Carol Becker, Jane Gaines, James Schamus, and Jana Wright at Columbia. Nic Sammond has been a patient auditor of my unformed ideas and gets his own sentence. Thanks, too, to Mark Lynn Anderson, Paul Babiak, Richard W. Bann, Hilde D'haeyere, Kathy Fuller-Seeley, Eric Hoyt, Steven Jacobs, Frank Kelleter, Richard Koszarski, Judith Yaross Lee, Jeff Menne, Ross Melnick, Tom Paulus, Joanna Rapf, and Yair Solan for helping me improve and develop my arguments.

None of what follows would have been possible without the archivists who enabled my research: Barbara Hall, Kristine Krueger, Jennifer Romero, and Faye Thompson at the Academy of Motion Pictures Arts and Sciences' Margaret Herrick Library; Patricia Prestinary at California State University Fullerton's Special Collections; Peer Ebbighausen and Roni Lubliner at NBC Universal

Archives; Steve Massa at the New York Public Library for the Performing Arts; Lauren Buisson and Julie Graham at UCLA's Special Collections; Ned Comstock at USC's Cinematic Arts Library; Sandra Garcia-Myers at USC's Warner Bros. Archives; and Amanda Stow at the University of Wyoming's American Heritage Center. That I have come to plan my holiday vacations as archive trips is testimony to the companionship I always find there.

I have also benefited from the work of undergraduate and graduate research assistants who kindly took time from their own studies to help me: thank you, Michael Kaminski, Carolyn Condon, and Wentao Ma. Mauri Sumén, film composer for the Kaurismäki brothers, was a graduate student at Columbia while I was finishing this book and provided the musical transcriptions and audio clips for chapter 4. Columbia University's film and video technicians, Peter Vaughan and Michael Cacioppo Belantara, created the movie clips when they weren't otherwise occupied showing me how to use a remote control.

Financial support for this project was provided by the Connaught Start-Up Fund at the University of Toronto, Columbia University's Faculty Research Allocation Program, the Social Sciences and Humanities Research Council of Canada, and the Visiting Researcher Stipend Program of UCLA Film and Television Library's Archive and Study Center. For the latter, Mark Quigley worked tirelessly in facilitating my stay at the archive and deserves particular thanks. The Open Access publication of this work was enabled by a donation to Columbia University's School of the Arts and a grant from the Columbia Open-Access Publication fund.

Two remarkable editors at the University of California Press navigated the book through its ten-year passage, Mary Francis and Raina Polivka. It was Mary who saw the possible value of my writing a "sequel" of sorts to my first monograph, *The Fun Factory* (2009); and it was Raina who, inheriting the project in its final stages, encouraged me to explore the parameters of Open Access publishing. Their associates, Aimée Goggins and Zuha Khan, patiently held my hand through the big little things that crowd any book's final stages.

My research and writing were happily delayed in 2012 by the birth of Samantha and Sullivan. That this occurred two months after my relocation to New York created a special kind of chaos out of which I now find myself, four years later, with a rather splendid family. My partner, Inie Park, gave up several years of weekends to let me hole up in my office while a real-life slapstick onslaught was happening just outside the office door. The twins gave me an invaluable education in humor, even though they'd choose *Dinosaur Train* over Laurel and Hardy every time. If there is something humming in the background of the following pages, it is my attempt to see my childhood self in you, kids, and to find my present self in my parents, Marilyn and Peter King, to whom this book is lovingly dedicated. The reasons will crop up occasionally in what follows.

*

Material from chapter 3 was initially incorporated into "'The Spice of the Program': Educational Pictures, Early Sound Slapstick, and the Small-Town Audience," *Film History* 23, no. 3 (2011): 313–330, and sections from chapter 5 are adapted from "Slapstick and Mis-Remembrance: Buster Keaton's Columbia Shorts," *New Review of Film and Television Studies* 5, no. 3 (December 2007): 333–352.

NOTE ON DATES

In what follows I cite the month and year of release of all short subjects, but only the years for features. In instances when an exact release date is impossible to ascertain—as, for instance, for many early Vitaphone shorts—I give a best guess as to the rough time frame when the film was first placed in distribution.

Introduction

Keyword: Hokum

The word "hokum" is one of several examples of stage slang whose meaning, at a certain point in the 1920s, was much debated. According to a 1926 article in *American Speech,* it was the "most discussed word in the entire vernacular" of popular entertainment (another was "jazz").[1] The term seems to have origins in the late nineteenth century, perhaps deriving from "oakum" (material used to calk the seams of a ship; by extension, "sure-fire" gags and other material used to secure the success of a stage act) or, alternatively, as a combination of "hocus-pocus" (sleight-of-hand, trickery) and "bunkum" (nonsense). Still, those origins are sufficiently questionable that novelist Edna Ferber, in her 1929 *Cimarron,* could claim that the term was of exclusively twentieth-century derivation. ("The slang words hokum and bunk were not then [1898] in use.")[2] The ambiguous sources of "hokum" also correspond to a split in its development, which, by the 1920s, had seen the sense of "sure-fire" shift in the more disparaging direction indicated by "bunkum." Writing in 1928, a reporter for the *New York Times* expressed incredulity that a term once describing material that "'get[s] over' . . . with an audience" was now synonymous with "hooey, tripe, apple-sauce, blah and bologna."[3]

The word seems to have something to do with comedy, although this is not invariable. An article in the *Times* of 1923 indicated a possible melodramatic reference as well, describing hokum as "old and sure-fire comedy. Also tear-inducing situations," which suggests hokum's applicability to anything that traded in strong or obvious effects, whether of comedy or of sentiment.[4] "Hokum is not always comedy; sometimes it borders on pathos" echoed the essay in *American Speech.*[5] Still, the reference to comedy, specifically of the knockabout, slapstick variety, was primary.

One early piece, from 1917, parsed the term generally as "low comedy verging on vulgarity."[6] Others were more specific, concentrating a retrospective sense of the term by referring it to residual traditions of comedic performance. "It is doubtful whether the most inveterate of theatergoers knows what is meant by the term 'hokum stuff,'" noted one writer in 1915, explaining, "It is an old-time minstrelman equivalent for slap-stick comedy."[7] A decade later, *Vanity Fair*'s Walter Winchell referenced circus clowning: "Actors who redden their faces, and wear ill-fitting apparel, and take falls to get laughs are 'hokum comics.'"[8] The *New York Times* meanwhile used cinematic examples, relating "hokum" to one- and two-reel slapstick shorts of the 1910s:

> When Charley [*sic*] Chaplin smeared somebody's face with a custard pie, that was considered good gag [*sic*]; but when every comedian of the one and two reels made use of the idea, then it became hokum.
>
> A considerable number of rarely humorous devices for laughter were invented by the old Keystone Comedy [*sic*]; and every once in awhile, some of these ancient tricks crop out [*sic*]. Then somebody acquainted with the true meaning of the word, cries "hokum!"[9]

It is difficult to read far in the flurry of these articles without perceiving in "hokum" the symptom of a shift in comic sensibility. The word had more than merely ambiguous meanings: it had an unmistakable trajectory that shifted from description (gags that "get over") to denigration ("old-time," "apple-sauce"). That trajectory, moreover, crested sharply around the mid-1920s, when the term was apparently never more widespread. (A Google Ngram search reveals that the word's frequency was highest in 1926, constituting 0.000023 percent of words in now-digitized US books, an over 15,000 percent increase from the start of the decade.)[10]

The later sense of "old-time" is not surprising: one of the characteristics of knockabout or slapstick comedy is that it has often been disparaged as passé, a disavowed yardstick ever since the movement in American variety theater toward polite vaudeville in the 1880s and 1890s. Yet the sudden popularization by the mid-1920s of a cant or slang term for that status bespeaks a more confident spirit of devaluation. In this sense the secret meaning of hokum's ascendancy is the decisive banalization of a comedic style that, in vaudeville as in film, had once formed a contested mainstay of early twentieth-century mass culture. This book will track the sources and processes of that devaluation as it unfolded in the years to come. My focus will fall squarely on film—already by the 1920s the primary venue where slapstick was encountered by the American public—and within that focus, I will be concentrating not so much on feature-length films as on the one- and two-reel subjects where, according to the *Times*, hokum was commonest currency. The introductory pages that follow flesh out my reasons for these choices and establish the historiographic premises that will underpin my investigation.

THE "END" OF SLAPSTICK? TWO PREMISES

Premise 1: Rethinking Sound

The idea that film slapstick sank into abrupt decline in the late 1920s may seem familiar. One of the hoariest clichés of comedy history holds that Hollywood's conversion to sound—beginning in 1926 and completed by 1929—profoundly changed the course of film comedy's development. The coming of sound, it is said, represents a decisive turning point at which the art of the great silent clowns—Charlie Chaplin, Buster Keaton, and others—came to an end, hinging instead into the crude realism of lesser talents like the Three Stooges. But it is precisely this sense of an "end" that we might first want to come to grips with here, since it will be part of my argument that film slapstick's troubled history from the late 1920s on has been misleadingly framed. Why, for instance, has the coming of sound commonly been thought of as a kind of Rubicon moment vis-à-vis screen comedy? Why is comedy, uniquely among film genres, so clearly divided into silent versus talkie eras? After all, as film historian David Kalat has suggested, there is no comparable discrimination that would mourn the end of the "silent western" as though technological change alone amounted to a decisive generic mutation.[11] With comedy, though, it is as if sound has come to constitute nothing less than an allegorical gap dividing screen comedy's Edenic glories from its subsequent Fall. Three classic accounts can serve as evidence.

James Agee's eloquent 1949 *Life* essay, "Comedy's Greatest Era," is perhaps the most celebrated of these, establishing many of the basic premises of this master narrative. Agee's essay was crucial in positioning the silent features of Chaplin, Keaton, Harry Langdon, and Harold Lloyd as a kind of Mount Rushmore of comic achievement, and it did so by using sound as a kind of whipping boy against which the performative virtuosity of the silent clowns might best be measured. "When a modern [i.e., sound] comedian gets hit on the head," Agee wrote,

> the most he is apt to do is look sleepy. When a silent comedian got hit on the head he seldom let it go so flatly. He realized a broad license, and a ruthless discipline within that license. It was his business to be as funny as possible physically, without the help or hindrance of words. So he gave us a figure of speech, or rather of vision, for loss of consciousness. In other words he gave us a poem, a kind of poem, moreover, that everybody understands. The least he might do was to straighten up stiff as a plank and fall over backward with such skill that his whole length seemed to slap the floor at the same instant. Or he might make a cadenza of it—look vague, smile like an angel, roll up his eyes, lace his fingers, thrust his hands palms downward as far as they would go, hunch his shoulders, rise on tiptoe, prance ecstatically in narrowing circles until, with tallow knees, he sank down the vortex of his dizziness to the floor, and there signified nirvana by kicking his heels twice, like a swimming frog.[12]

But such pantomimic virtuosity simply did not lend itself to dialogue, which, in Agee's opinion, belonged to an entirely separate performative tradition. "Because [the motion picture now] talks, the only comedians who ever mastered the screen cannot work, for they cannot combine their comic style with talk."[13] Agee's is, in this sense, a kind of technologically determined history of performative practice: cinema's silence demanded a newly expressive form of physical performance—one requiring the "talents of a dancer, acrobat, clown and mime"—that could not "combine" with dialogue humor.

This basic sense of incompatibility would carry through a quarter century later into the first scholarly overview of film comedy, Gerald Mast's 1973 *The Comic Mind*. Once again, a technological limitation (silence) is said to have created a performative form that simply could not survive the altered climate of talking pictures. With sound, we are told, a "particular kind of comedy died." "One of the reasons great physical comedians developed in the teens and twenties was that the potential of the medium demanded their services. The physical comedian who communicated personality, social attitudes and human relationships by physical means . . . was an outgrowth of a medium whose only tools were movement, rhythm, and physical objects and surfaces."[14] Mast's conclusion from this sounds perplexing—"Sound comedy is structural, not physical"—but by this he means simply to convey the idea that sound comedy is more structure*d* insofar as it is "carefully molded in advance" by the director and writer. If the essential relation in sound comedy is between the writer and the page, Mast proposes, then the kernel of silent comedy was the bond between the clown's body and the camera.[15]

The idea that silent comedy was *not* "carefully molded" at the scripting stage is completely spurious, of course, at least as concerns the overwhelming majority of slapstick shorts and features from around the mid-1910s on. Yet despite these vagaries, Mast does introduce an important addition to the Agee template by premising his argument on an explicitly stated theory of filmic art, which a brief footnote attributes to Rudolph Arnheim's classic 1933 study, *Film as Art*. As Mast glosses Arnheim: "It has often been said that art is a function of limitations, that the province of art is precisely that gap between nature and the way nature can be imitated in the work of art." Hence, Mast concludes, film comedy became art to the degree to which comedians found expressive physical means to "compensate for [the] gap" that the medium's silence had installed.[16]

It is this position that critic Walter Kerr would develop two years later, in his magisterial 1975 study *The Silent Clowns*. Arnheim is no longer cited, but Kerr's language makes the indebtedness unmistakable. "Logically, art begins in a taking away," Kerr argued. "Each limitation on the camera's power to reproduce reality . . . [paved] the way to an exercise of art."[17] (Compare the wording of Arnheim's 1933 text, which argues that cinematic artistry depends on "robbing the real event of something"—on withholding attributes of color, three-dimensionality, and sound—such that silent film therefore "derives definite artistic possibilities from its silence.")[18] For screen comedy, this distance from reality became the foundation of the form's silent-era achievement as fantasy: the appeal of the silent clowns, Kerr argued, rested in their liberation from the laws of the ordinary physical world. "None of the limitations of the silent screen . . . seemed limitations to its comedians. Rather, they seemed opportunities for slipping ever more elaborately through the cogs of the cosmic machinery, escaping the indignities of a dimensional, hostile universe. Fly through the transom when a policeman locks the door? Why not?"[19] Sound, in restoring the realism of the filmed image, killed the fantasy on which comic artistry depended.

> [Sound] gives the lie to the very kind of comedy—the very original kind of comedy—audiences had cherished: comedy in which the real world had not only been tamed but, in its dreamlike submission and in its swift unexpected conjunctions, made lyric. . . . With the form itself gone, we could no longer see [the silent comedians] as we had once seen them: as mysteriously mute archetypes who had made bizarre bargains with a half-imagined, half-authentic world. The game was over.[20]

We will have cause to unpack these models further over the course of this book. For the present, it is important only to note how their prioritization of technological factors produces, in each case, a historiography grounded in discontinuity. The technological properties of cinema—its original silence and subsequent voice—are approached as decisive mutations in comedic representation: the former generates a new performative tradition as an "outgrowth" (Mast) of the medium, while the latter stops it dead. What we get by way of explanation is thus little more than the "random autonomy of invention" from which other forms of causation (cultural, economic, etc.) are excluded from the outset.[21] What we might better pursue is an approach that, without discounting the genuine difference that technological changes make, nonetheless understands those changes as what media theorist Ian Hutchby calls "affordances" that frame—but do not inevitably prefigure—possible directions of development.[22] Only when we stop confusing a technological property with a comedic tradition will it become possible to conjoin screen slapstick before *and* after the conversion era as a developing trajectory whose historical explanation exceeds a merely technological determinism.

Premise 2: Displacing Features

But it is perhaps not only the reification of silence that is a problem for the history of screen comedy. Historical apprehension of slapstick's varied fortunes into the sound era has also been obstructed by a focus on the individual careers of the canonized slapstick clowns. To the extent that film historians have prioritized the careers of Chaplin, Keaton, Langdon, and others, they have fractured the history of slapstick into a series of incommensurable narratives. Buster Keaton's career, we learn, was derailed by a coincidence of personal and professional misfires that were his alone: the loss of artistic control when he signed with MGM in 1928; the deterioration of his marriage to Natalie Talmadge; his growing alcoholism. Harry Langdon fell victim to hubris in firing his director, a young Frank Capra, and opting to direct himself in three disastrously received subsequent features—*Three's a Crowd* (1927), *The Chaser* (1928), and *Heart Trouble* (1928)—that torpedoed his career at the moment of the industry's transition. Charles Chaplin meanwhile became an outlier many times over, in part through artistic choice (alone among his peers, he refused to capitulate to the talkie trend until he finally allowed the tramp to talk in *The Great Dictator* [1940]), in part by inactivity (he completed only two features in the 1930s, *City Lights* [1931] and *Modern Times* [1936]), in part by his socialist political commitments. My point here is not to question the at times quite dubious accuracy of these narratives (the story of Langdon's "hubris," for instance, originates in Capra's autobiography and is hardly disinterested).[23] Rather, I am concerned with the ways these explanatory models displace any apprehension of the larger province of slapstick filmmaking within which these comedians worked: a historiography oriented around the great silent-era clowns inevitably collapses into the irreducible singularities of so many careers. It is, perhaps, the very incommensurability of the great comedians' passage through these years that gives a spurious legitimacy to the one thing they all shared—the transition to sound—as a master explanation.

This problem commends a shift of focus to short subjects. Such a shift would not only, by definition, pull the historiography of American film comedy out from the shadows of the famed feature-length comedians, but it would also restore a fuller sense of film slapstick's place within what sociologist Pierre Bourdieu would have described as the "field" of comedy production during this period. The notion of a field, as developed primarily in Bourdieu's 1992 *The Rules of Art,* designates in the most straightforward sense the "social microcosm"—the relations and interactions among individuals, groups, and institutions—that makes up a given sphere of cultural production (so that one may speak, for instance, of the "literary field," the "intellectual field," and so forth).[24] I will take up later the question of how field analysis opens onto surrounding social formations; for the present, I only want to note how receptive the short-subject sector is to such an approach, allowing for a more complex inventorying of the variables and controversies involved in slapstick's sound-era decline. The world in which short-subject comedians and their filmmakers moved was not simply a miniaturized enclave of the film industry; it was a sphere of filmmaking shaped by a sui generis network of interpersonal, professional, and institutional relations radiating out of the studio gates into the larger fields of vaudeville, burlesque, circus, and even literary humor within which hierarchies of comedic value—and slapstick's place within them—were defined. Sam Warner, of Warner Bros., recognized as much when he assigned to ex-vaudevillian Bryan ("Brynie") Foy the initial responsibility of managing the earliest Vitaphone shorts, until 1931: formerly one of the Seven Little Foys, Brynie's résumé made him uniquely suited to call in the talents of the nation's best vaudeville acts, in the process shaping a house style at Warner of "virtual Broadway."[25] Professional networks were also defined by prior affiliation within the film industry, often reaching back many years. A case in point is provided by the short-subject unit at Columbia Pictures under the stewardship of Jules White, first appointed to the studio in 1932 by Columbia boss Harry Cohn. "He [Cohn] never bothered me," White recalled. "He didn't consider my department important enough to bother with"—a hands-off policy that allowed White, together with his brother Jack, to chase up professional connections from their years in silent comedy and build a roster of veteran comic talents.[26] Outside of above-the-line talent, meanwhile, the social microcosm of short-format comedy was also shaped by new technical and musical personnel brought into the industry during the transition to sound: the Hal Roach Studios' 1928 contract with the Victor Talking Machine Company, for instance, would bring to the studio a number of Victor employees—sound engineer Elmer Raguse, A&R man Leroy Shield—who would play major roles in innovating the soundscape of early talkie comedies. In sum, if we seriously want to understand the transformations in sound-era slapstick, we need to patiently enumerate the full range of different agencies and actors that mediated these transformations; we need, that is, to "follow the natives," instead of abstracting a handful of "artists" (Chaplin, Keaton, etc.) or mobilizing a few global causes (sound, modernity, etc.) to which are attributed a mass of effects.[27] The short-subject sector, qua field, permits just such an analysis.

Also favoring shorts is the very direct optic they provide on audience taste and demand. Not only was the short subject the most widely disseminated filmic format for the slapstick idiom—the major studios typically produced and/or distributed between two and three dozen one- and two-reel comic shorts every year—but it was also the only one whose function was framed in purely representative terms. It was simply as comedy that each slapstick short was promoted by exhibitors (common tags in newspaper ads were "plus two-reel comedy" or "news, comedy, cartoon"), and it was simply as comedy that each was assessed. "Here's a comedy that is a comedy," "Not [Charley] Chase's best comedy, but still a comedy," "Sold to us for a comedy, but is poor."[28] The seeming circularity of such evaluations bespeaks the absence of any presumption that short-subject comedy might be evaluated outside of the simple criterion of funniness: the value of short-subject comedy was simply that it should "be" comedy, with no expectation of surplus, of, say, artistic experiment or idiosyncratic deviation. The very transparency of these expectations indicates the short subject as a streak-free window onto changing sensibilities: if short-subject slapstick "declined" during the 1930s, then this was surely in part because the style of comedy in question no longer answered to audiences' entertainment needs, at least within the industry's primary markets.

The same conclusion follows from the peculiarities of short-subject distribution during this period. One of the distinctive aspects of the short-subject sector, in comparison with features, was the greater flexibility for exhibitors to select from each studio's offerings, based on their predictions or understandings of audience demand. An early case in point was provided by Warner Bros., which by the summer of 1928 yielded control over the choice of shorts to the exhibitors who had contracted for its services, offering a catalog that included newer titles alongside previous releases. As film historian Charles Wolfe notes, "The Vitaphone shorts were [thus] treated less as motion-picture events than as a commercial library of recorded performances . . . that could be rented and replayed on an ongoing basis" and, presumably, selected according to the tastes of local audiences.[29] Other studios did not go quite that far but, at least after 1933, permitted varying degrees of exhibitor choice. In that year, the practice of what was called "full-line forcing"—a controversial extension of block booking, whereby the major studios had forced theater owners to take their full lines of shorts as a condition of accepting features—was outlawed under the auspices of the National Industrial Recovery Act (NIRA). Whereas previously exhibitors had often been forced to accept many more major-studio short subjects than they could possibly play in a single season—a strategy designed to freeze out independently produced shorts—the NIRA's Motion Picture Code was more equitable: distributors could now force shorts only in proportion to the number of rented features.[30] Exhibitors' subsequent freedom of selection can be illustrated by the case of the Interstate circuit in Texas, which drew interested trade commentary when it took the unique step of establishing its own short-subject booking department in 1934: the department's five-person team—headed by the appropriately named Besa Short—would preview short subjects, assemble them into programs for the circuit's different theaters, and arrange subsequent bookings based on audience feedback, practices that remained in place at Interstate for the rest of the decade. "No other film property has the latitude and elasticity in booking as has the short subject," Short proclaimed.[31] It is in fact this very margin of elasticity that renders shorts so receptive an interface for examining patterns of exhibitor need during this period. Again, if short-subject slapstick declined in the sound era, then a plausible hypothesis would surely posit a lack of primary-market demand. We will want, further, to examine this in terms of distribution: where, if anywhere, were these films reliably booked? and where no longer?—questions I take up in chapter 3.

The final justification for short-oriented history is the most straightforward: historiographic neglect. Perhaps no area of American film comedy has been so entirely overlooked as the field of the slapstick short subsequent to the coming of sound. To the extent that the sound-era short has drawn any scholarly attention, it has been not as a mainstay of the American clown tradition but as a laboratory for working through the textual and technological practices of the sound film more generally: Warner Bros.'s pioneering sound-on-disc Vitaphone shorts thus loom large in conventional histories of the coming of sound, as do Walt Disney's early experiments with sound-image relations in cartoons like *Steamboat Willie* (November 1928); more recently, scholars like Jennifer Fleger and Katherine

Spring have explored the role of musical shorts in carving out an aesthetic identity for film sound.[32] There is also, in the existing literature, a familiar defense of the short subject during this period, one that defers to their important role within the "balanced program" concept of the era's exhibition practices. Short subjects, this line of argument goes, had value not only as a necessary "buffer" to the feature presentation, but also in ensuring the diversity of appeal necessary to sustain a mass audience. Travelogues, cartoons, slapstick, and sing-alongs all constituted just a fraction of the many and varied genres of shorts during this period, to say nothing of the more outré examples, featuring, for example, talking dogs (MGM's *Dogville* comedies, 1930–1931), golf instruction (Vitaphone's *How I Play Golf, by Bobby Jones* shorts, 1931–1932), and glee club recitals (Educational's *Spirit of the Campus* series, 1932–1933), among many others. Still, none of this quite gets us to the specific historicity of the short subject qua short subject; that is, the changing parameters—industrial, economic, textual, and so on—that shaped short-format filmmaking during this period. Nor, again, does it really have anything to say about the comedies that constituted so sizable a portion of the short-subject field. The latter omission is particularly startling given the historic importance tradition-ally ceded to slapstick two-reelers of the *silent* era—for instance, Charlie Chaplin's Mutual releases (1916–1917), Roscoe Arbuckle's Comique two-reelers (1917–1920), or Buster Keaton's Metro/First National shorts (1920–1923). By comparison, the names of short-subject comedians from the early sound period—Clark and McCullough, Andy Clyde, Edgar Kennedy, and Thelma Todd, among many oth-ers—testify to a largely forgotten history. Even the Three Stooges, despite their lon-gevity as slapstick's leading practitioners in two-reel talking pictures (1934–1958), have drawn next to no academic interest.[33] The slapstick short has in this sense been a structuring absence in the historiography of early sound Hollywood; this book aims to rectify that.

THE SOCIOLOGY OF RESIDUAL CULTURAL FORMS

This monograph is a follow-up to the historiographic project commenced in my previous book, *The Fun Factory: The Keystone Film Company and the Emergence of Mass Culture* (2009). There, looking closely at the Keystone Film Company (1912–1917), I examined how slapstick, as a "low" cultural form with roots in plebe-ian and working-class subcultures, was transformed during the 1910s into a mass-cultural cinematic genre with cross-class appeal. Keystone's popularity, I argued, was characteristic of the cultural and commercial energies of the period, which saw a proliferation of cheap commercial entertainment forms that outstripped their original audience to forge a new "mass" cultural orientation. Slapstick was just such a form—a "lively art," to borrow the label coined by critic Gilbert Seldes to describe forms like jazz, comic strips, and vaudeville comedy—and its popularization bespoke an era in which cultural boundaries were being redrawn by the mongrel energies of modern mass culture. In this sequel, by contrast, I explore the reverse image of that process, tracking the constitutive processes that troubled and eventually undermined slapstick's mass appeal, precipitating its decline into cultural banality during the years of the Depression.

In this sense, *Hokum!* is a study of how once-dominant cultural forms become "residual" in the precise sense suggested by Raymond Williams. Any culture, Williams tells us, includes "available elements of its past" that nonetheless remain "active in the cultural process." A residual element is one that no longer retains its former efficacy but nonetheless remains sufficiently "live" that it will in most cases have to be incorporated if the new dominant culture is to retain any coherence; which is to say that it will have to be reinterpreted or resignified—idealized, diluted, reified, what have you—so that its continued presence does not register as a threat. (One of Williams's examples is the idea of rural community, whose potential challenge to the values of urban industrial capitalism is blunted by its very idealization as pastoral nostalgia.)[34] But what is it that leads cultural forms toward the residual? For Williams, the answer is found in terms of the standard Marxist categories of class analysis: cultural forms pass into residuality when the social formations that produce them are displaced by the "formation of a new class, the coming to consciousness of a new class."[35] Applied to the case in hand—the changing fortunes of film slapstick—Williams's hypothesis provides a valuable starting point: changes in comic sensibility beginning in the 1920s can indeed be linked to changes in the class nature of American society, specifically the emergence of what Paula Fass calls the "peer society" of 1920s urban life—a perspective that will be developed in my first chapter.[36] But, in and of itself, Williams's master key cannot account for how, for instance, these new class configurations were themselves energized by and swept up in a broader metropolitanism that in fact outstripped the boundaries of class; nor does it secure a perspective on film history that would avoid the one-to-one reductionism of reading industry determinants directly back into social formation. We have not gained much if we simply replace a *technological* determinism in terms of the coming of sound with a no less simplifying *social* determinism in terms of class.

Again, Bourdieu can help us here. The concept of a "field of production" implies the imperative of analyzing the structure of that field at a number of levels, both "internal" and "external," that resist the positing of any single determinism in the last instance. As the sociologist explains (he is talking about literature):

> The science of cultural works presupposes three operations which are as necessary and necessarily linked as the three levels of social reality that they apprehend. First, one must analyse the position of the literary (etc.) field within the field of power, and its evolution in time. Second, one must analyse the internal structure of the literary (etc.) field, a universe obeying its own laws of functioning and transformation,

meaning the structure of objective relations between positions occupied by individuals and groups placed in a situation of competition for legitimacy. And finally, the analysis involves the genesis of the habitus of occupants of these positions, that is, the systems of dispositions which, being the product of a social trajectory and of a position within the literary (etc.) field, find in this position a more or less favourable opportunity to be realized.[37]

This formulation seems to me to have the advantage of comprehending a given field in terms of the social processes in which it is enmeshed (in terms, e.g., of its position within a given field of power, in terms, too, of the social dispositions and backgrounds of its "occupants"), even as it allows the literary field an autonomy ("its own laws of functioning") that exceeds any direct or unmediated anchoring in external determinants. The concept of a field of production thus usefully allows for forms of determination that are both "internal" and "external": the former, as noted, covers the "structure of objective relations between positions occupied by individuals and groups" *within* a given field, including its own hierarchies of power and legitimacy; the latter addresses the social trajectories that bring individuals *into* the field, as well, I would add, as the field's interface with its surrounding public/audience. With respect to the short-subject industry in the years following sound, the internal dimension will, then, cover such areas as the complex economic and institutional realities negotiated by short-subject producers within the broader film industry (yes, the transition to sound, but also factors like the major studios' growing involvement in short-subject production/distribution, the advent of double bills and changing exhibition practices, etc.), as well as the "competitions for legitimacy" among comedy producers; the external dimension meanwhile encompasses the backgrounds and dispositions of short-subject filmmakers themselves, as well as the processes that continually shaped and reshaped the constitution of slapstick's audience.

Within this framework, I would like to posit three key "moments" in slapstick's passage to cultural residuality. (This list is not meant to be a broadly applicable model but simply covers those processes emerging from the present study.) The first—what I will call *rehierarchization*—addresses the way formal and stylistic innovations within a given field of cultural production serve as catalysts for establishing new distinctions within that field, so that formerly "dominant" forms become demoted and passé. The innovation of sound, I will argue, served as a catalyst in just this latter sense. Its introduction spurred a kind of land rush on the part of short-comedy filmmakers to explore the possibilities of what producer Al Christie labeled the "new style" of dialogue comedy, with the result that slapstick came to occupy the contrasting role as the "old."[38] The second moment is what Bourdieu has theorized as *banalization*, which refers to the ways in which devaluation is inseparable from social change within a given form's audience. What counts here is the way a formerly dominant cultural product comes to be considered *déclassé* when it finds a new audience among a devalued social group—a process I trace through the marketing of slapstick to small-town audiences during the years of the Depression.[39] These first two moments concern the various *positional* shifts—internal and external, respectively—that govern residuality within a given field: (a) the changing position of a cultural form vis-à-vis the hierarchy of other forms, (b) its changing position vis-à-vis the hierarchy of possible publics. (These two dimensions are strictly correlative.) By contrast, the third addresses the *affective* logic that invests residual forms with a kind of surplus value and, as such, renders their residuality not only safe (like the pastoral idealization of rural life) but also profitable. Within the mass cultural marketplace, that process has long been fulfilled via the operations of nostalgia, of which an early example, I will suggest, is evidenced in the "old-time" slapstick craze of the mid- to late 1930s. Nostalgia here stood for the affective logic whereby the "old" could be reinvested as "old-time," the outdated reclaimed as throwback; it was the very process through which slapstick as a residual form nonetheless sustained a lingering place and economic function within the mass cultural market.

At this point, the skeletal framework of my argument is starting to come into focus. One last methodological point, however, deserves brief mention before I turn to a more formal chapter-by-chapter overview. I have pointed a few times to the notion of mass culture as the broader framework for my analysis, by which is meant the interconnected culture industries that, for much of the twentieth century, orchestrated the production of cultural goods on a rationalized, assembly-line basis, guided by the dictates of the market. In my earlier study of Keystone, I sought to demonstrate the constitutive hybridity of mass cultural forms, citing Max Weber to the effect that the marketplace overrides cultural distinctions by requiring the producers of cultural goods to fuse genres and cross boundaries to achieve the broadest spectrum of appeal.[40] I would now like to supplement this by reading mass culture more explicitly in terms of its function of *managing difference*—that is, of organizing diverse publics into imaginary associations configured around textual forms and the modes of their circulation.[41] The media's "mass" functioning, in this sense, depends upon modes of discursive address that negotiate social divisions which might otherwise hinder this circulation; moreover, particular divisions (of, say, class, gender, race, or region) will, at particular times, become more or less prominent within the industry's market operations in response to processes of social change. The history of mass culture might then be thought of in terms of the various "differences" that, in any given period, are prioritized within these modes of address and configured into the imagined entity of a "mass" public (with the caveat that the terms of this configuration can always be contested by those who refuse the place thereby assigned them, as we will repeatedly see in what follows). Slapstick's ascendancy in the

1910s, for example, was predicated on its ability to perform this role in relation to the class differences that provided early Hollywood with arguably its most prominent dichotomy (this is the one-sentence version of *The Fun Factory*). The hypothesis, then, will be that a form becomes residual when changes in those dichotomies disable this function, consigning the form in question to one or the other side of a new set of orchestrating divisions. Slapstick's decline in this sense bespeaks changing hierarchies of comedic value that were no longer primarily governed by the dichotomies of class that spawned the form's initial success (which will turn out to be the one-sentence version of *Hokum!*—but please keep reading).

CHAPTER BREAKDOWN

My argument proceeds through five chapters, divided into two parts. Part 1 establishes contexts—both the cultural context of new directions in comic sensibility from the late 1920s on (chapter 1) and the film industrial context of the marketplace for short subjects (chapter 2), each of which secured the slapstick short's progressive obsolescence in the decade after sound. The first chapter accordingly begins by marking a paradox for contemporary scholars who have interpreted early twentieth-century slapstick as a quintessentially "modern" comedic form; namely that, by the end of the 1920s, film slapstick's modernity was already significantly qualified, the form increasingly perceived as outdated, as evidenced by a series of disappointing box-office showings for prestige slapstick features like Charlie Chaplin's *The Circus* (1928), Buster Keaton's *The General* (1927), Harry Langdon's *Three's a Crowd* (1927), and Harold Lloyd's *The Kid Brother* (1927), all of which drew receipts far lower than the comedians' previous work. Further, the very years that witnessed these comedians' first notable falterings also saw the emergence of a new critical perspective associating modern humor not with film slapstick at all, but with a new vein of metropolitan absurdism exemplified by the city wits who staffed publisher Harold Ross's *New Yorker* as well as by a new cohort of "cuckoo" comedians like the Marx Brothers and Joe Cook who began to dominate the Broadway scene. The effort to define this vogue means that my argument starts, paradoxically, by pushing short subjects—indeed, film in general—to the back burner in order to establish the larger cultural coordinates that shaped new directions in comic sensibility. Older class-based hierarchies of cultural value, I show, were being recast during this period in terms of cultural geography, inaugurating a dichotomy between the urbane cultural vanguard and small-town hokum that will resonate throughout this book. The opening chapter strikes these themes by first examining the era's argot of comedic "lunacy" and "goofyism" as a new vocabulary of metropolitan distinction, before turning to short subjects for a case study of the comic style of Bobby Clark and Paul McCullough, the duo that best enshrined this cuckoo mode in shorts, first at Fox (1928–1929) and subsequently at RKO (1930–1935).

The second chapter turns more singularly to the film industry, to offer a synoptic overview of the short-subject industry's main lines of development in the decade following sound. My focus here falls in part on a number of significant challenges faced by the short-subject sector during this period: the advent of sound, of course, as well as the trend of the double bill, which drastically restricted the scope for shorts on theater schedules. More centrally, however, I am concerned with the changing role of shorts within the industry's evolving understanding of its public, pre- and post-Depression. In the earliest years of sound, the Hollywood studios had characteristically addressed their public in terms of a new language of metropolitan distinction: Hollywood initially conceived of sound cinema as a means of cultural dissemination and uplift bringing the Broadway vanguard to a nationwide audience divided by geographic and cultural distance. Consumer resistance to these marketing strategies, particularly in the heartland, soon provoked a change of tack, however, as the industry next began to seek a newly populist appeal informed by New Deal–era ideals of civic inclusivity. The rhetoric of cultural distinction and hierarchy thus yielded to one of public service as modes of audience address. This chapter shows how these two modes flanked the period of this study, where they were enshrined in the competing market strategies adopted, first, by Warner Bros. for the launch of its pioneering Vitaphone sound shorts beginning in 1926 and, second, almost a decade later, by MGM's revamped short-subject unit under Jack Chertok. The short-subject comedies of Algonquin wit Robert Benchley—first at Fox (1928–1929), later at Chertok's MGM unit (1935–1940, 1943–1944)—will provide the culminating case study of the book's first part, to illustrate how these alternative frameworks of address were articulated through comedy.

Part 2 consists of three chapter-length case studies of individual short-subject producer/distributors, each designed to yield a richer understanding of the determinants and forms of what I thematize as slapstick's "social aging" during this period, its passage toward the déclassé or out of date.

The third chapter explores the history of Educational Pictures (slogan: "The Spice of the Program") from the transition to sound to the company's decline in the late 1930s. Established in 1915 by Earle W. Hammons, Educational had, by the end of the silent era, become the industry leader in short-comedy distribution, serving over thirteen thousand exhibitors with a regular program featuring comedians Larry Semon, Lloyd Hamilton, Charley Bowers, and others. Yet, within five years of the transition to sound, the company's reputation had sunk precipitously, its sound shorts notorious as a bargain-basement home for aging comedians. In assessing the implications of that decline, I focus on how industry developments squeezed the company's output out of major urban markets and so underwrote slapstick's assumed affiliation with the "naïve" tastes of hinterland publics during this period. Industrial marginalization was thus conflated with cultural devaluation as Hammons's organization now had little choice but to reorient its output for those selfsame publics. Notable here was an upsurge in rural comedies (especially with the ascendant popularity of "hick" comic characters at Educational, like Andy Clyde, Harry Gribbon, and even Buster Keaton), as well as a growing incorporation of hillbilly and southern music traditions into the company's films. As such, moreover, Educational's fate exemplifies the operations of banalization as a mode of the social aging of cultural forms—that is, the way certain cultural practices (in this instance, slapstick) become outmoded through a process of social change in their audience.[42]

For the fourth chapter, the focus shifts to the Hal Roach Studios and the distinctive role that music played in negotiating Roach's passage through the upheavals of the early sound era. In focusing on the studio's experiments with musical formats for comedy, this chapter aligns itself in part with emerging scholarly trends focused on the convergence of film and recorded sound industries wrought by the introduction of electrical sound technology to cinema.[43] New corporate relations with the Victor Talking Machine Company brought Roach talented musical personnel and technicians who innovated new wall-to-wall (that is, continual) scoring practices that drew upon modern, Tin Pan Alley–style jazz idioms. But these experiments with slapstick musicality would take a quite different turn a few years later, when Roach's efforts to leverage his company toward the production of features resulted in a spate of feature-length Viennese-style operettas starring Laurel and Hardy in period costumes (*The Devil's Brother*, 1933; *Babes in Toyland*, 1934; *The Bohemian Girl*, 1936). In embracing operetta as a format, Roach's filmmakers were not only elaborating on the pioneering musical tendencies evident in their earliest sound shorts; they were also cautiously opting for the security of middlebrow "family" appeal to mollify the financial risks of moving into features. Excised from the contemporaneity of vernacular idioms—both comedic (slapstick) and musical (jazz)—Laurel and Hardy were now conscripted to what Susan Stewart theorizes as the "infinite time" of fairy tale.[44] What was cemented was thus a second mode of slapstick's social aging, the resignification of the clown no longer according to the lumpen typology of turn-of-the-century vaudeville and early film but as a pantomime-like figure for childhood reverie.

The final chapter turns to the aforementioned old-time slapstick vogue of the late 1930s and the role of nostalgia in "re-membering" slapstick's meaning and function as a residual form. The term "re-membering" I derive from sociologist Barbara Myerhoff, for whom it refers to a type of nostalgia that seeks the "reaggregation of [a group's] members, the figures who belong to one's life story."[45] By the end of the 1930s, I suggest, Hollywood was gripped by a similar project of nostalgic investment, as it sought to respond to widespread criticism that it had lost touch with its audience. Within these efforts, moreover, silent comedy came to serve as an emblem for the industry's "gay and goofy" past in whose image Hollywood now sought to reimagine itself. The focus of my analysis here falls primarily on the output of the Columbia short-subject department, which, beginning in 1932, was reorganized under the supervising team of Zion Myers and Jules White. Of all short-subject firms, Columbia's seems to have been most oriented toward the throwback market, a haven of sorts for out-of-work slapstick veterans who, together again, assembled a pastiche style that restored the knockabout energies of earlier Mack Sennett/Keystone–era comedy—most famously in the rough-and-tumble farces of the Three Stooges (1934–1958). This chapter reconstructs that comedic restoration. But it also seeks, finally, a *political* vector to slapstick's outdatedness by assessing the relationship linking popular cultural forms (like slapstick) to populism. At a surface level, the nostalgic restoration of 1910s-vintage slapstick in the context of the 1930s makes a certain cultural sense: both were populist eras, the latter perhaps forcing memory to return fondly to the cultural forms of the former. But this could not be accomplished without abstracting from the very different political conceptions that animated those earlier forms: put simply, the *class-conflictual* populist style that Columbia inherited from Sennett was not the same as the *civic-inclusive* populism that came to characterize New Deal America's political rhetoric, resulting in strange contortions of comedic formulas whenever the studio's filmmakers sought to engage the present. The Depression-era retrofitting of Sennett-style farce as nostalgia only confirmed that the form's moment as a "live" vehicle of social representation had long since passed—the third and final trajectory of the form's social aging.

*

Rather than an official conclusion, *Hokum!* closes with a coda, which tracks the ongoing rewriting of slapstick as nostalgia in subsequent decades. The market for "old-time" comedy that first opened in the 1930s proved to be a geyser that continued to spout reissues for many years to come, for instance, in the cycle of vintage-comedy anthologies that began to take off in the late 1950s (e.g., the numerous Robert Youngson–produced compilation films, *When Comedy Was King, Days of Thrills and Laughter,* etc.) as well as in the recycling of early comedy shorts in syndicated children's television programming (e.g., *The Funny Manns* and *Fractured Flickers*) beginning in the early 1960s.[46] This is where historiography and autobiography become inseparable for me, for it is here that my own connection with slapstick was first made. I have indelible memories of the compilation TV show *Harold Lloyd's World of Comedy,* produced by Time-Life for PBS in the early 1970s and a staple of early evening programming in Britain (where I'm from) a decade later. After the kids' shows had finished on the main channels and my mother was cooking fish fingers, I would switch to BBC2 and wait for the memorable theme song ("Hooray for Harold Lloyd / doo-doo doo-doo doo-doo doo-doo doo doo!"). In evoking this affective bond, I claim no singularity. The historiography of slapstick cinema has long been stained with nostalgia: few writers on the form have failed to include some kind of personal reminiscence.[47] And while it might be true that nostalgia is not history—that it "cannot replace the difficult task of reconstructing and interpreting the past"—there is something to be said for a historicizing of nostalgia itself and an accounting of its sources.[48] This finally is what *Hokum!* seeks.

PART ONE

Contexts

"The Cuckoo School"

Humor and Metropolitan Culture in 1920s America

April Fool's Day, 1923: The Palace Theatre at 47th and Broadway is hosting National Vaudeville Artists' Week, a fundraiser for the Keith-Albee Organization performers' union. On the bill are such luminaries as singer Sophie Tucker, Ben Bernie and His Orchestra, musical comedy two-act Herb Williams and Hilda Wolfus, contortionists the Luster Brothers, "nut" comedians Sam Dody and Sam Lewis, and others.[1] It is Dody and Lewis, we will imagine, who will make a particular impression on one member of the audience that night.

No description of Lewis and Dody's performance that night has been located. But we know from extant reports and phonograph recordings that their act from this period typically consisted of a single song—"Hello, Hello, Hello!"— extrapolated to include a variable number of stanzas, each one ending with the triple salutation and interrupted with absurdist asides. They come onto the stage, lifelessly grunt a gibberish refrain, introduce themselves with deadpan affect— "I'm Mike," says one, "I'm Ike," the other—and launch into a verse, delivered less as a song than as a matter of fact.

> Just the other night
> Right near from here,
> We saw a funny sight:
> A couple they were dispossessed;
> The wife stood there in tears.
> That's the first time they'd been out together in twenty years.
> Hello, Hello, Hello!
> Hello, Hello, Hello!
> You can't milk a herring.
> Hello, Hello, Hello!

Sticking to their toneless voices, they reintroduce themselves "("I'm Mike," "I'm Ike") and commence another verse:

> We filed our income tax.
> And there is where
> They pull such funny cracks.
> They asked us "What's a post office?"
> We answered there and then,
> "That's the place that Scotchmen go to fill up fountain pens."
> Hello, Hello, Hello!
> Hello, Hello, Hello!
> Fish don't perspire.
> Hello, Hello, Hello!

Without any apparent interest in their own performance, the two continue this way for the rest of the act, adding stanza upon stanza in an apparent effort to show, as one reviewer put it, that "the limit in the number of extra verses to a song has not yet been reached," before abruptly ending, as flatly as when they started.[2] "[One more] final statement of complete nonsense," one writer put it, "wholly incongruous, uncalled for, beside the point where there is no point"; one more affectless "Hello, Hello, Hello!" and they retire.[3] The audience is beside itself with laughter.

Let us now cast as our impressed theatergoer none other than Gilbert Seldes, the iconoclastic former editor of the New York–based cultural journal *The Dial*, who was then in the process of writing his pioneering *The 7 Lively Arts* (1924). We do not know with certainty whether he attended that night, but we do know that (1) he saw Lewis and Dody perform their "Hello, Hello, Hello!" routine at some point prior to the publication of *The 7 Lively Arts* exactly a year later, (2) the National Vaudeville Artists' show was the duo's last New York appearance until August of the following year, again at the Palace, and (3) Seldes was an inveterate devotee of vaudeville, so it is no stretch to place him there—if not exactly on the April 1 opening night, then likely at some point that week.[4] We also know that the essays he was preparing for *The 7 Lively Arts* include some of the most important writing on popular comedy, stage and screen, from this period. Seldes's effort in that landmark book was to bring what he called the "lively arts"—including slapstick film, comic strips, and vaudeville—into the same field of criticism as the canonized arts. Not only was there no intrinsic "opposition between the great and the lively arts," he contended there, but also the latter were "more interesting to the adult cultivated intelligence than most of the things which [today] pass for art"—a thesis that Seldes would expound upon in now well-known chapters on vaudeville performers Fanny Brice and Al Jolson, on the slapstick films of Charlie Chaplin and the Keystone Film

Company, on George Herriman's Krazy Kat ("Easily the most . . . satisfactory work of art produced in America"), among other topics.[5] The popular arts, Seldes believed, demanded recognition as genuinely vernacular forms whose achievements had been obscured by the "bogus" aesthetic values of the "genteel tradition."[6]

But the Seldes who perhaps attended the Palace that April night did so not merely as an influential proselytizer, but also in search of new directions. In later years, he would characterize the 1920s as the "last flowering of our popular arts," but the worm was already in the bud in the essays he was then preparing.[7] The 7 Lively Arts piece on Keystone is a case in point: nothing in cinema's early development, Seldes averred, had been more suited to the medium's expressive means than Keystone-style slapstick, whose "freedom of fancy . . . [and] roaring, destructive, careless energy" could "appear nowhere if not on the screen."[8] Yet the form had of late been overtaken by the "poison of culture."[9] The selfsame genteel tradition that had obscured recognition of the popular arts had installed itself in the creative ambitions of their practitioners. "Its directors have heard abuse and sly remarks about custard pies so long that they have begun to believe in them, and the madness which is a monstrous sanity in the movie comedy is likely to die out."[10] Nor was this a stand-alone example. Throughout the book, Seldes seems caught in the dilemma of expressing enthusiasm for cultural objects that repeatedly disappoint his image of them. As though trying to plug a leaky dam, he celebrates the Follies-style revue but chastises those of its directors who introduce "element[s] of artistic bunk"; he qualifies his encomium of Jolson by dismissing the "second-rate sentiment" of his Mammy songs; he favors vaudeville for its "damned effrontery" of genteel tradition while in almost the same breath lamenting its "corrupt desire to be refined," and so on.[11] The Gilbert Seldes who encountered Lewis and Dody circa April Fool's Day, 1923, was a critic primed to place faith in a comedic duo whose very uniqueness promised to break this cycle of compromise.

And so it was that the team came to occupy a key position in an essay that Seldes next began to prepare, not for 7 Lively Arts, but for Vanity Fair, in which he identified a new current in American comedy. The essay's title and the current it describes were one and the same: "The Cuckoo School of Humour in America," published in May of 1924, one month after his book's release (fig. 1).[12] Describing a "certain madness" in contemporary comicality, Seldes there made Lewis and Dody the central and longest example in a litany of modern absurdism extending through the diverse work of vaudevillians Joe Cook and Ed Wynn, Algonquin litterateurs Robert Benchley and Donald Ogden Stewart, songwriting team Bert Kalmar and Harry Ruby, and sports journalists and humorists Arthur "Bugs" Baer and Ring Lardner, and exemplified in the "equivocal nonsense" of song titles like "Yes, We Have No Bananas" and "When It's Night-Time in Italy, It's Wednesday

46   VANITY FAIR

The Cuckoo School of Humour in America

A Not-Quite-Profound Analysis of the Increasingly Dada Strain in the Cleverness of Our Fun-Makers

By VIVIAN SHAW

AT the present moment, no musical show in New York is complete unless it possesses the following joke:

A: Didn't I meet you in Buffalo?
B: I never was in Buffalo in my life.
A: Neither was I. Must have been two other fellows.

Of course, a great deal depends on the manner of saying the last few words. They can be said apologetically, or casually, or as if illuminating a puzzling question. In most cases, the result is funny. Of the four thousand men and women now employing the joke in order to earn a living, probably not more than two thousand are aware of its psychological basis, or know that it has been used in text-books of philosophy to illustrate some particularly provoking dilemma. It is really a sort of extended pun, which, as Santayana says, suspends the fancy between two incompatible but irresistible meanings.

"The meaning doesn't matter if it's only idle chatter", says Bunthorne, in the Gilbert and Sullivan opera; and that idea seems to be gaining ground with the American arts. A certain madness—or, if you prefer, irresponsibility—is the mark of one mode of humour at the moment; and that it isn't new is obvious, because it makes you think at once of *Alice of Wonderland*. In that book, there is a remorse-

ARTHUR "BUGS" BAER

The creator of "The Family Album" in the Hearst Sunday newspapers is an American humourist whose style recalls nothing so much as the humour of Falstaff's companions. His comparisons are wildly exaggerated; his metaphor is brilliant and effective. He always leaves out the expected, and is so fantastic, so irresponsible, and so funny as to be a veritable master of "cuckoo humour"

into a funny stanza. Not very funny at that. And, at the end, they say their hellos again and utter a final statement of complete nonsense, wholly incongruous, uncalled for, beside the point where there is no point. This final point is as follows: "Worms have no expression, or fishes don't perspire; Hello Hello Hello."

The two singers are extremely good in handling this cuckoo material; they stick to an almost toneless voice; one says, "I'm Mike"; the other, "I'm Ike", without the faintest show of interest in the proceedings. It's all rather like reading a rhymed statement of fact. And it is incredibly funny. Here is a specimen stanza of a song, as intoned by them:

And just the other night
Right near from here,
We saw a funny sight:
A couple they were dispossessed;
The wife stood there in tears—
That's the first time they'd been out
Together in twenty years.
Hello, Hello, Hello!
Hello, Hello, Hello!
Make mine vanilla,
Hello, Hello, Hello!

This is so good a stunt in vaudeville that it has been promptly worked into a revue. In *Kid Boots*, Mr. Ziegfeld's latest musical revue, the material is elaborated and a quartette sings

FIGURE 1. Vivian Shaw, "The Cuckoo School of Humour in America," *Vanity Fair*, May 1924.

Over Here," as well as in what he identified as a vogue for "shaggy" jokes. It is with such a joke that the essay begins:

> At the present moment, no musical show in New York is complete unless it possesses the following joke:
>
> A: Didn't I meet you in Buffalo?
> B: I never was in Buffalo in my life.
> A: Neither was I. Must have been two other fellows.

Of course, a great deal depends on the manner of saying the last few words. They can be said apologetically, or casually, or as if illuminating a puzzling question. In most cases, the result is funny. Of the four thousand men and women now employing the joke in order to earn a living, probably not more than two thousand are aware of its psychological basis, or know that it has been used in text-books of philosophy to illustrate some particularly provoking dilemma. It is really a sort of extended pun, which, as Santayana says, suspends the fancy between two incompatible but irresistible meanings. . . .

A certain madness—or, if you prefer, irresponsibility—is the mark of one mode of humour at the moment; and that it isn't new is obvious, because it makes you think at once of *Alice of Wonderland* [*sic*]. In that book, there is a remorseless logic; but it is a logic which exists in another world: a logic of the fourth dimension.[13]

But it is in Lewis and Dody that the purest "delight of nonsense" is to be found, and Seldes spends a full four paragraphs describing their act. "I do not recall any previous practitioners of such songs [as 'Hello, Hello, Hello!'], nor have any rivals achieved the combination of apparent dullness with insanity," he concludes. "In any case, the type of nonsense is now associated with the name of Lewis and Dody," who, by Seldes's reckoning, had already spawned "about half a dozen" imitators.[14]

Written under the pseudonym of Vivian Shaw, the essay was no doubt something of a test balloon: the diverse examples of cuckoo humor contravene Seldes's endeavor, in *The 7 Lively Arts,* to corral comedic innovation exclusively to the side of the popular arts, even as the essay also brackets off cinema as a source of these new directions. For one thing, what Seldes meant by the "cuckoo school" was primarily a verbal or literary style—manifest as a "nonsensical juxtaposition of words"—which, as such, left silent motion pictures unsurprisingly untouched. True, "comedy plots [in film] are occasionally ludicrous, and once in a while you get entertaining visualizations of 'looking daggers' or 'seeing stars'"—Seldes seems to be thinking here of director Del Lord's fondness for animated effects in his work for Sennett around this time—"but these are not quite the same thing."[15] For another, the advent of the cuckoo style represented a shift in the cultural vectors of comedy. Neither vernacular nor genteel, the new style of "irresponsible madness" provoked Seldes instead to draw analogies with contemporary modernism, pointing to "the increasingly Dada strain in the cleverness of our fun-makers" and "an unintentional Dadaism."[16]

We will have cause to return to Seldes's assessment of the cuckoo school shortly. For the present, it is enough to note how his assessment bespeaks not simply the beginnings of a shift in the direction of American humor but also an emerging transformation in the idea and location of the "modern" in respect to comedy. The popular-festive mode of Keystone-style slapstick that so impressed critics who chafed against the constraints of "great art" was, Seldes feared, being reclaimed to those same constraints; in turn, the laurels of comedic modernity—once claimed by slapstick as a form whose "galvanic gestures and movements were creating fresh lines and interesting angles"—were being relinquished to the newly "Dadaist" spirit of the cuckoo wits.[17] Nor was Seldes the only observer to catch the change in the comedic winds. Critics on the lookout for emerging directions in comicality in the 1920s generally equated the new with the nonsensical. This was a period when commentators began enthusiastically inventing terms like the "New Nonsense," the "Larger Lunacy," the "Higher Goofyism," or "Humor Gone Nuts," in Donald Ogden Stewart's felicitous phrase, to describe a style whose "moonstruck quality," another wrote, rested on "utterly fey *non sequiturs*" and "puns of monstrous absurdity."[18] But slapstick no longer excited such enthusiasms. The theater critic George Jean Nathan, who had described Chaplin's *A Dog's Life* (April 1918) as "tremendously superior at almost every point to much of the vulgar low comedy of such as Shakespeare," could barely stifle a yawn a decade later at the same comedian's now "inelastic technique" and overreliance on the sentimental themes "of a bygone day."[19]

All of which will likely come as news to those scholars who, in recent years, have elevated silent-era slapstick film as a vanguard of cinematic modernity. Without differentiation as to period, scholars have understood silent slapstick as a genre linked to the rhythms of "the piston [and] the automobile," providing an "anarchic supplement" to the systematization of modern mass culture.[20] Slapstick's modernity has in this way been cast in the image of a kind of inverted Fordism, as a carnivalesque reflex against assembly-line routinization and technocratic rationalization whose trajectory extended from the simplest trick devices of early gag films (hoses, pieces of string, jerry-rigged umbrellas) to the "uproarious inventions" of Sennett-style slapstick and beyond.[21] What is more, these readings have been buttressed by the insights of European avant-gardists and cultural critics of the period—the famed Frankfurt School, in particular—whose scholarly popularization has informed a consensus understanding of slapstick as a preeminent cultural form for negotiating, through laughter, the pressures of modern subjectivity.[22] "One has to hand this to the Americans," the German critic Siegfried Kracauer wrote in 1926. "With slapstick films they have created a form that offers a counterweight to their reality: if in that reality they subject the world to an often unbearable discipline, the film in turn dismantles this self-imposed order quite forcefully."[23] But no mention is made here of the fact that, by the mid-1920s, the front lines of comedic innovation had moved elsewhere, and a curtain is drawn on the chorus of complaints that the era's leading slapstick clowns—Chaplin, Buster Keaton, Harry Langdon, and Harold Lloyd—were out of step with critical and public tastes. If, then, in what follows, I choose to dissent from these influential Frankfurt School readings, this is less on the assumption that they were in some absolute sense "wrong" in their assessment of slapstick's cultural resonance; rather, in a more relativist spirit, it is to propose that slapstick's *claim* on modernity—its claim to a laughter that was, if one may say so, of its moment—was being displaced and disavowed within comedic hierarchies of the 1920s.[24] From this perspective even an imagined conjunction—that chiasmus of critical reevaluation that may or may not have passed through the figure of Gilbert Seldes on April Fool's Day, 1923—may hold clues to a major comedic reconfiguration whose forms would filter across the mediascape of the 1920s, as this chapter will explore.

"TO BE URBAN BY BIRTH AND PREFERENCE": METROPOLITAN CULTURE AND THE MODERNIZING OF MIRTH

We begin with a basic taxonomy of comic creativity: the division between the clown and the wit, which I want to use as a way of sketching the changing social contexts of humor production during this period. Etymologically, "clown" signifies a "rustic booby," a simpleton, the reverse of the savvy townsperson who is the "wit."[25] There are several distinctions embedded in this binary; simplicity versus sophistication, physical horseplay versus droll wordplay, country versus city, and so on. And an initial point to be made is that a major current of comedic innovation in the 1920s entailed the displacement of the clown and the ascendancy of the urbane wit as an engine of comedic renewal. "[Until 1895, our] chief humorists long had been rustic or western," argue Walter Blair and Hamlin Hall in a seminal study of literary humor. "Many of them during the years 1895 to 1915 were country-born but became city dwellers. Beginning about 1915, our most famous humorous writers were to be urban by birth and preference."[26] The result, as critic Carl van Doren pointed out in 1923, was a clustering of "town wits"—"licensed jesters of the town" who "promulgate the jests, discuss the personalities, [and] represent the manners of New York"—including critics and columnists like Franklin P. Adams, Robert Benchley, Dorothy Parker, and Alexander Woollcott, to whom new periodicals like *Vanity Fair* (founded in 1913), *College Humor* (1921), and the *New Yorker* (1924) lent a platform.[27]

Such a shift stands as an obvious reflex of urbanization, as the corollary of an era in which, as the 1920 census had shown, the rural-urban balance tilted decisively toward the latter. But it also bespeaks the role that humor came to play in giving definition to new patterns of class sensibility and distinction within the era's city life. Sociologist Paula Fass has spoken in this connection of the "peer society," a constellation of middle-class, college-educated city folk who came of age with the nation's urban culture.[28] This peer society was, in the first instance, a symptom of the growing role of schools and colleges in industrial society: whereas previous generations of children had been nurtured into family-based value systems (in turn inflected by multigenerational practices of ethnic and class belonging), the generation of the 1920s grew up in an environment dominated by peers, with whom they spent increasing time in a variety of formal and informal school- and college-based activities. Between 1900 and 1930, Fass notes, college enrollments tripled and high-school enrollments increased more than sixfold; the result, she argues, was to secure a transition from family membership to the performance of social roles as the framework within which middle-class identity was articulated.[29] The appeal of city living, in this context, was to provide a setting within which these new, extrafamilial modes of identification could be continued after graduation: city life—its mores and its eccentricities—became a kind of second campus for middle-class urbanites who learned to couch their class privilege in the language of urbane sophistication. Peer connections gave humor periodicals like the *New Yorker* the "flavor of an alumni reunion" among Ivy-educated writers, while the magazine's signature "Talk of the Town" section gave readers access to a clubby perspective on the local scene.[30] *College Humor* similarly boasted of its exclusive appeal to a young, college-educated, and urban readership ("We start with the premise that our current readers are educated"), while the *New Yorker* flagged its urbane credentials in continuous disparaging references to the "little old lady in Dubuque."[31]

The discriminatory impulses that spurred these wits to throw barbs outside the metropole also led them to repudiate the regional humor traditions that had flourished there. It was in such a spirit, for example, that Harvard-educated Bostonian and then-*Life* drama critic Robert Benchley published "The Brow-Elevation in Humor" in 1922, in which he disparaged earlier traditions of regional dialect humor (when it was "considered good form to spoof . . . surplus learning of any kind" and when "any one who wanted to qualify as a humorist had to be able to mispronounce any word of over three syllables"), advocating instead a literate approach that would address itself to, rather than ridicule, an educated mindset.[32] The journalist Franklin P. Adams, who published under the acronym F.P.A., was on this count the superior of Mark Twain, Benchley concluded. (F.P.A. himself similarly used Twain as a whipping boy for measuring the superiority of the new, urbane set, claiming sports journalist and essayist Ring Lardner to be a keener observer than Twain and a "better hater of the human four-flusher as he is.")[33]

The modernizing impulse that took shape from this process of comedic position taking has been characterized in various ways.[34] For literary scholar Ann Douglas, the *New Yorker* style of humor shared in the broader metropolitan sensibility of what she calls "terrible honesty," an irreverent hostility to older sentimental pieties that became a generational style for American moderns. "Think of Dorothy Parker publicly skewering a young man who announced he 'could not bear fools,'" Douglas writes. "'That's funny,' [Parker] replied, 'your mother could.'"[35] For others, the peer society found its distinctive voice in the genre of "little man" humor, tales of the quotidian distresses of city life in which *New Yorker* writers and city columnists excelled.[36] The humor of the peer wits in this sense has come to imply at least two complementary dynamics: on the one hand, a gesture of urbane differentiation, enshrined in a terse and cultured style that abjured the ethical imperatives of Victorian gentility even as it rebelled against the studied ingenuousness of regional humor; on the other, a gesture of club-house affiliation, in the commitment to experiences (little man humor) and values (Benchley's "signs of culture") in which middle-class city folk recognized their commonalities.

Yet it would not be altogether correct to comprehend this new humor purely in terms of these characteristic distributions of tone and material; also relevant are the formal processes of wit in which the new metropolitan cohort's self-conscious spirit of renewal became concrete. Consider, for example, the famous "optimist" joke that ran repeatedly in early issues of the *New Yorker,* a two-line joke told backward ("A man who can make it in par." "What's an optimist, Pop?").[37] Typically understood as editor and founder Harold Ross's attack on well-worn humor conventions, the joke exemplifies a style of logico-linguistic nonsense beloved of the urbane wits. What many of these humorists shared—and what in fact united them with the broader cuckoo style of "equivocal nonsense" identified by Seldes—

was a proclivity for what semiotician Paolo Virno has dubbed the "entrepreneurial joke." Characteristic of entrepreneurial joking, Virno avers, is the formal impulse to deploy or rearrange semantic resources in new and absurdist ways, to recombine or disorder elements of linguistic praxis in a fashion that stretches beyond—or even outright contravenes—ordinary codifications.[38] Cameo examples of the technique are found in the enumeration of semantic items that do not "fit" within a given category (a favored technique of Robert Benchley, as in the quip from his 1921 essay "All About the Silesian Problem," where we learn that the Silesian *Bienen* government consists of three classes: "The nobles, the welterweights, and the licensed pilots"), in the nonsensical variation upon the same verbal material (as in Donald Ogden Stewart's 1928 "How We Introduced the Budget System into Our Home," which closes with a line that redeploys the term "budget" in an illegitimate way: "Any budget caught smoking will be instantly expelled"), as well as in a penchant for metaphorical overreach (as in the adventuresome comparisons of sports writer "Bugs" Baer, who wrote of a knocked-down fighter that "his face looked like a slateful of wrong answers").[39] We also see this semantic playfulness, in an immediate way, in the titles of Benchley's literary humor collections from this period, such as *No Poems, or Around the World Backwards and Sideways* (1928), *From Bed to Worse, or Comforting Thoughts about the Bison* (1934), or, a title meant to confuse booksellers, *20,000 Leagues under the Sea, or David Copperfield* (1928). Frequent *New Yorker* commentator Frank Sullivan meanwhile provided what amounted to a metacommentary on this style in his 1928 essay "Should Admirals Shave?" which takes semantic disorder as its explicit theme. An investigation into the reasons behind the putative "loose talk" of navy admiralty unwarrantedly petitioning for war, Sullivan's essay uncovers the cause to be the "imprisoned words" caught in the admirals' bushy beards:

> "These imprisoned words," [I explained to the admiral,] "have remained in your beard indefinitely. A word is very tenacious of life. It lives for years; for centuries, especially when nourished in the cozy warmth of a beard. Now, you must also realize that all these accumulated words have mingled with the newly-arriving words as each new speech coursed through your beard. . . . So that your speeches have really been hash."
>
> "Hash?"
>
> "Yes, hash. Portions of old speeches mixed indiscriminately with your new speeches. That's why they made no sense." . . .
>
> "By George," he exclaimed, "I never thought of that."
>
> "By George who?"
>
> He thought that over for a moment.
>
> "By George, I guess you've got me there," he said. "I don't know what George I say 'By George' about."
>
> "Sloppy diction," I rebuked. "Lazy mental processes. Don't do it, my good man. Settle upon one George and stick to him."
>
> "I choose George Herman Ruth [i.e., "Babe" Ruth]," said the admiral.[40]

It was accordingly in terms of a conflict between the sense and the forms of language that new directions in American humor asserted themselves, not without controversy. "Philosophically, our modern American humor is utterly nihilistic," was the dismayed opinion of MIT English professor Robert Rogers, in a lecture given in 1932. "Logic has no validity, the laws of cause and effect are a washout. The world is an *Alice in Wonderland* world"—the Carroll reference, again—"where even logic is dream logic and everything is unexpected." "This kind of humor cannot be written by plain people," Rogers added, in a telling acknowledgment of the peer society's class orientation. "It is the fruit of a rather extreme type of modern education, plus a considerable knowledge of modern art and science."[41] The Canadian humorist Stephen Leacock expressed similar reservations—despite being something of a hero to the *New Yorker* circle of writers—when he disparaged their predilection for nonsensical wordplay as mere "mechanics": "Humor resting on words alone is only for the nursery, the schoolroom and for odd moments. . . . It reaches its real ground when it becomes the humor of situation and character."[42]

Leacock's advocacy of "situation and character" betrayed the continued, if waning, sway of older, genteel theorists of laughter like George Meredith and W. M. Thackeray, who had advanced an ethical-humanist vision of humor in keeping with Victorian ideals of self-discipline and moral improvement.[43] Yet there were other literary observers who proved more willing to follow new trends in reappraising the value of nonsense. Perhaps the most notable intervention in favor of the lunatic vogue came from New York radical intellectual Max Eastman, whose 1936 study, *Enjoyment of Laughter,* constitutes one of the only fully formed theories of laughter to emerge from this period in American letters. A spirited, if ultimately rather quibbling, rejoinder to Sigmund Freud's "release" theory in *Jokes and Their Relation to the Unconscious* (1905), Eastman's study posits childhood pleasure in nonsense not, as with Freud, as a psychogenetic stage *prior* to humor—to which state humorous expression seeks to restore us— but rather as the latter's essential seed; put another way, he reads the toddler's uninhibited pleasure in verbal play as humor's origin and not its destination. "Freud's great sin against humor, and against the art of enjoying it, is that he makes it all furtive. He thinks, as we have seen, that there is no humor in the playful nonsense of children, and that humor arises only when grown-up people elude their ideals of rationality and other inhibitions."[44] By contrast: "The first law of humor is that things can be funny only when we are in fun," and that "'being in fun' is a condition most natural to childhood"—a vantage from which Eastman reevaluates nonsense not as "pre-" humorous but as humor's purest expression.[45] Eastman's leading examples of nonsense? Lewis Carroll (again), *New Yorker* writers Donald Ogden Stewart, E. B. White, and James Thurber, and Broadway comedian Joe Cook. Eastman's commentary on Cook's 1930 humor book, *Why I Will Not Imitate Four Hawaiians* (also the name of the comedian's most famous routine), exemplifies his understanding of nonsense as a restoration of childhood play.[46]

> Let us read a bit of Joe Cook's *Why I Will Not Imitate Four Hawaiians*, that book which he talked off impromptu, in the presence of a stenographer, much, I imagine, as a child talks nonsense to himself but aided by the presence of his elders.

> THE STORY OF MY LIFE—SO HELP ME

> First of all, I am worth millions of dollars. The fact that I was born in a private family proves that my parents must have been well-to-do. My tender age, at the time of birth, created quite a bit of favorable comment. I was the only child in kindergarten that chewed tobacco. The parents of all the other little boys and girls begged me to teach their children this delightful habit. Gertrude Ederle was the first woman to swim the English Channel. . . .

> Does the statement, "Gertrude Ederle was the first woman to swim the English Channel," contribute anything of . . . relevance to Joe Cook's *Life—So Help Me*[?]. . . I asked the author why he put that phrase in, and he said: "It just came into my head and I saw that it was funny—I don't know why?"[47]

In keeping with the operative procedures of cuckoo wit, linguistic matter is here displaced from the register of signification to that of form: the Ederle line has nothing, in terms of meaning, to commend its humor beyond the purely formal and syntactic disconnect with what precedes it.

There was, to be sure, a relation with literary and artistic modernism here that has not gone unrecognized. Scholars of American humor have long acknowledged the presence of e. e. cummings behind Don Marquis's *archy and mehitabel* tales, written in the voice of a cockroach (archy) who is the reincarnation of a writer of lower-case free verse. More recently, literary scholar Sanford Pinsker has noted the affinities between T. S. Eliot and James Thurber and between James Joyce and S. J. Perelman.[48] Still, this was an uneasy relationship, in which literary modernism excited the nonsensical proclivities of the New York wits less in a spirit of homage than in a kind of bemused travesty—what historian Leonard Diepeveen usefully dubs "mock modernism."[49] Don Marquis was perhaps the columnist to have worked most consistently in this parodic vein, not only in the *archy and mehitabel* series from his "Sun Dial" column, but also in stand-alone pieces in which he contributed doggerel verse commentary on the poets of the hour, such as the early piece "The Golden Group" from 1915:

> "Peppercorns and purple sleet
> Enwrap me round from heat to feet,
> Wrap me around and make me thine,"
> Said Amy Lowell to Gertrude Stein.

> "Buzz-saws, buzzards, curds and glue
> Show me my affinity for you.

You are my golden sister-soul,"
Said Gertrude Stein to Amy Lowell. . . .

And Hermione harked to them, rapt, elate:
"And aren't they all of them simply great!
Of course, the bourgeois can't understand—
But aren't they wonderful? Aren't they grand!"[50]

But others of the peer society's esteemed wits also contributed to the genre—including F.P.A., Thurber, and White—each of whom found in artistic modernism a provocation to heighten their ludic play with linguistic form, only now in the direction of satire.[51] It is in fact in such mock modernism that we see cuckoo humor's paradoxical relation to the orbit of modernist reinvention: what the era's leading wits and tastemakers commonly embraced in the field of comedy they just as frequently condemned in the field of serious literature. The same Max Eastman who approvingly cited Joe Cook's non sequiturs in his *Enjoyment of Laughter* did not extend similar courtesy to modernist literature in his 1931 study, *The Literary Mind*. Developing a position more famously struck in his well-known 1929 *Harper's* essay "The Cult of Unintelligibility," Eastman there pretended to a sympathetic analysis of a passage of what he termed "Gertrudian prose": "I was looking at you, the sweet boy that does not want sweet soap. Neatness of feet do not win feet, but feet win the neatness of men. Run does not run west but west runs east. I like west strawberries best." After some serious discussion, the kicker comes when Eastman reveals the passage to be not authentic Stein but "the ravings of a maniac cited by Kraepelin in his Clinical Psychiatry."[52] "Every one who has composed poems knows how often he has to sacrifice a value that is both clear and dear to him, in order to communicate his poem to others," Eastman summarizes. "Abandon that motive, the limitation it imposes, and you will find yourself writing modernist poetry"—or, one could add, cuckoo humor.[53]

The further risk in charting this link with literary modernism, however, is that it makes the cuckoo style appear too rarefied a field, so it is important to recall that the tendency toward this style of linguistic and logical play cycled widely, through both "sophisticated" and "popular" arts. Seldes's foundational "cuckoo" essay was perspicacious in this regard: the author's deep-seated commitment to popular forms made him uniquely attuned to the expansive ambit of contemporary nonsense, not merely as the argot of Algonquin humorists but as part of a new metropolitan vernacular that filtered through the New York humor scene. Nor, moreover, should we deduce from this a "trickle-down" dissemination—as though peer society wits like Robert Benchley originated a style that was subsequently enshrined in popular ditties like "Yes, We Have No Bananas." What was occurring, rather, was a mutation of cultural hierarchy in which New York's newfound preeminence overflowed the conduits of older class cultures. Newly fascinated by their own resources, American moderns were precipitating what Ann Douglas describes as a "shift in cultural power from New England to New York"; from "reform-oriented, serious-minded, middlebrow religious and intellectual discourse to the lighthearted, streetwise, and more or less secular popular and mass arts as America's chosen means of self-expression."[54] In such a climate, cuckoo humor was hardly end-stopped in the performance of the peer society's cohesion but enjambed with discursive registers that informed the topsy-turvy heteroglossia of metropolitan life in other ways—in the double entendres and coded play of the nascent pansy craze, in the chaotic subversions of language of Jewish Broadway stars like the Marx Brothers (about whom more to come), even in the racial polyphony of jazz itself.[55] We need to think of the 1920s as a decade in which former class-based divisions pitting "genteel" against "popular" cultural spheres were being reorchestrated in terms of cultural geography, remapped onto new divisions separating the metropolis from the hinterland.[56] What the peer wits achieved was less the wholesale invention of a new style than its adaptation to changing coordinates of distinction: the "new nonsense" became, in their hands, one of the whetstones against which the contours of metropolitan distinctiveness were sharpened. Our next task will be to revisit our opening taxonomy to ask where this revolution in the nation's *wit* left the critical evaluation of some of the nation's most recognized *clowns*, on both stage and screen. On whom, in short, did the light of metropolitan favor still shine?

"MOST OF THE CRITICS HAVE HAD A HAND IN WRITING THE SHOW": CRITICS, COLLABORATORS, AND 1920S BROADWAY COMEDY

A first answer: not the slapstick comedians of feature-film fame. It is not well enough recognized, for instance, that all of the since canonized clowns of the silent era—Chaplin, Keaton, Lloyd, and Langdon—lost a step circa 1927–1928. All of them made expensive and boundary-pushing films during this period, respectively, *The Circus* (1928), *The General* (1927), *The Kid Brother* (1927), and *Three's a Crowd* (1927), and all four garnered bemused reception and disappointing box office.[57] *The General,* for example, was Keaton's most ambitious feature, a period piece which, at a price of nearly a million dollars, was the costliest film the comedian ever made. Although Keaton scholars have debated whether it made that money back, less debatable is the critical pasting the film received. "The fun is not exactly plentiful," Mordaunt Hall wrote in the *New York Times.* "Here [Keaton] is more the acrobat than the clown."[58] Analogous complaints were lodged against *Three's a Crowd,* Langdon's first attempt to direct himself and a box-office disaster. "For some reason or other Mr. Langdon has gone intensely tragic," wrote the critic of the *New York World.* "He appears to have forgotten for the time all he has ever learned of the value of movement and life in the making of comic pictures."[59] Chaplin's *The Circus* meanwhile may have done considerably better business (almost two million in domestic rentals by the end of 1931: he was Chaplin, after all) but in retrospect is recognized as a troubled production that the comedian preferred to forget.[60] Marketed as "a low-brow comedy for the highbrows," the film also marked a subtle shift in the comedian's critical evaluation.[61] Reservations were now openly voiced about the very pathos that had won the comedian his early acclaim. "Slapstick, by now, had become 'highbrow'; and Chaplin's pathos had been much praised," biographer Theodor Huff claimed. "So it is possible he overdid [it]."[62] The film effectively marked the end of Chaplin's contemporaneity. Finally, Lloyd's *The Kid Brother* fell unexpectedly flat right out of the gate, opening to widespread audience disinterest almost everywhere except New York and Los Angeles. As per Keaton's *The General, The Kid Brother* was an expensive, southern-set period comedy (a comedic reworking of the 1921 melodrama *Tol'able David*); as per Chaplin with *The Circus*, Lloyd was so crestfallen that he consigned the film to obscurity, even omitting it from his later compilation picture, *Harold Lloyd's World of Comedy* (1962). The point here is not to measure the merits of these films on the basis of their initial box office, but simply to note the general perception of the time that the most visible keepers of the "roaring, destructive" slapstick flame of the 1910s had blown it out.

The examples of Chaplin, Lloyd, and the "intensely tragic" Langdon are most instructive here. In appealing to pathos and sentimental melodrama, all three clowns were locked into standards of genteel appreciation that had fueled slapstick's growing acclaim in the 1910s—Chaplin's in particular—only subsequently to be foresworn by metropolitan critics of the 1920s.[63] "Charlie Chaplin has been gravely injured by reading discussions . . . of his tragic overtones" was thus the opinion of noted *New York Telegram* columnist Heywood Broun in 1928, while theater critic George Jean Nathan, as we have seen, disparaged those same overtones as the outmoded baggage of a "bygone day."[64] Nor was it only among urbane tastemakers that these comedians were now losing ground. Metropolitan critics who abjured an older sentimentalism as a block to modernity were, for a brief but crucial moment, in wholehearted agreement with hinterland publics who had long scorned the same as pretentious imposture. Such, at any rate, was the clear suggestion of a 1927 *Picture Play* essay, stridently titled "We—Want—Hokum!," which spoke up for small-town audiences whose preferences for the "good old days of slapstick" had been betrayed by the aspirations of the form's leading practitioner.

> The baggy trousers, the wisp of a cane, the stunted mustache and the monstrous shoes used to induce nothing short of hysterics in the great mass of the American public when Charlie was appearing in lowly two-reelers. But then he read someplace that the high art of his comedy lay in his ability to bring tears to the eyes. So he went

in for eight reels of throwing pathos, when the bigger half of his public would much rather see him hurl custard pies.[65]

The Circus was, then, hardly the first of Chaplin's films to falter with his public: the author of the *Picture Play* piece cited the evidence of numerous "small-town exhibitors who got a big bill but small patronage out of running 'The Gold Rush,'" including a Michigan exhibitor who "asserted that his customers paid to see Charlie in comedy, not drama" and an Arkansas theater owner who declared that Chaplin's "idea of comedy and my patrons' idea aren't the same."[66] The most acclaimed slapstick artist of his generation was suddenly losing credibility on both sides of a newly emergent divide pitting the "high-hatted" metropolis against small-town "hokum."

Nor was there much hope for metropolitan tastemakers in returning to slapstick's earthier sources. The plaudits formerly showered on burlesque low comedy in the 1910s by vanguard intellectuals like Nathan (who, in 1919, celebrated burlesque as the only true expression of "national humor") were plausible only as nostalgia by the late 1920s, as burlesque theater began its shuffle into unmistakable decline.[67] Changes in the economic structure of burlesque were forcing the tradition into banality. Prior to the 1920s, most burlesque companies operated as touring shows on circuits ("wheels," they were called): a performer could get a whole season out of a small set of new sketches and routines, performed each week in a different city and to a different audience. But by the mid-1920s, this business paradigm was displaced by "stock" burlesque: the theaters that once would have booked touring shows now found it cheaper to hire their own in-house companies and stage their own productions. Necessity thus became the mother of repetition: formulaic sketches and borrowed routines were an inescapable crutch for comedians who now had to come up with six to eight skits on a weekly basis. "There was nothing wrong with a new comedy scene," burlesque historian Rowland Barber pointed out in *The Night They Raided Minsky's*. "But it couldn't be *invented*. It had to be patched together out of bits of old bits from old burlesque shows."[68] Of course, popular theatrical forms like burlesque had always relied on this kind of "passed along" comedy material, but the shift to stock seems to have turned familiar comforts into critically irredeemable staleness. Seldes himself, writing in 1932, confessed to being dismayed by the "appalling monotony" that had by then settled on the form.[69] Even within their own domain, burlesque comedians now had to vie with the yawns and rustling newspapers of audience members (mostly working-class men) who were there primarily for the striptease dancers and scantily clad chorus girls.

Faced with the deterioration of the cultural reference points in whose name they had first sought to *épater le bourgeois,* taste-making critics relocated their enthusiasms to contemporary performers in whose work they perceived a kind of reflexive or baroque engagement with the legacy of variety past. In Broadway comedians like Ed Wynn (in his persona as the "perfect fool") or Joe Cook (the "one-man vaudeville show"), critics of 1920 perceived a kind of "innovative nostalgia" that not only corresponded to their own deep awareness of variety tradition but also excited their entrepreneurial spirit of cultural reinvention.[70] The best-known example is unquestionably the Marx Brothers, who, it is crucial to insist, were first and foremost a metropolitan phenomenon of the Broadway stage in the 1920s before they were a mass phenomenon on the nation's movie screens in the 1930s. Already by the time of their breakout success in the 1924 revue *I'll Say She Is!* the brothers had perfected stage personas as a kind of reflexive bricolage of the accrued weight of vaudeville convention—Chico's blending of novelty piano playing with an Italian immigrant routine, Groucho's acerbic one-liners and eccentric song and dance, Harpo's conflation of mute pantomime with the red-headed costuming of the stage Irish. For the New York critics who first encountered them in *I'll Say She Is!,* the Brothers thus impressed initially as a kind of zany palimpsest of comedic archetypes.[71] Writing in the *New York Herald,* critic Percy Hammond celebrated Groucho as a "nifty composition of all the humorous clowns, from William Collier down or up, as your taste may suggest."[72] George Jean Nathan struck a similar note: "The Marxes stem directly from [Harry] Watson, [George] Bickel and [Ed Lee] Wrothe and the various other comic teams that adorned the burlesque stage thirty years ago when it was at its zenith."[73] Algonquinite Alexander Woollcott meanwhile focused on Harpo— with whom he fell in love—placing him within the lineage of "that greater family which includes Joe Jackson and Bert Melrose and the Fratilini [*sic*] brothers, who fall over one another in so obliging a fashion at the Cirque Medrano in Paris."[74]

More significant, though, than the terms of their initial reception on Broadway is how critical admiration spawned creative collaborations harnessing the team to the contemporary absurdist vogue. As comedy historian Frank Krutnik notes, the success that the Marxes enjoyed as stars of Broadway—and which they subsequently carried over to Hollywood—"owe[d] a great deal to the alliance between these veteran vaudevillians and an elite group of New York intellectuals."[75] An unofficial patronage system was emerging in which the era's critics and litterateurs not only bestowed critical approval upon but also entered into creative partnership with those comedians who best answered to their tastes. The success of *I'll Say She Is!* resulted in a series of teamings with the city's leading literary humorists and critics, including the *New York Times* drama critic George S. Kaufman (who cowrote the Marxes' follow-up Broadway shows, *The Cocoanuts* [1925] and *Animal Crackers* [1928], both of which were adapted for the team's debut films, in 1929 and 1930 respectively) and, subsequently, *New Yorker* humorist S. J. Perelman (who provided the team with their first original film scripts, for *Monkey Business* [1931] and *Horse Feathers* [1932]).[76] While their breakout success with *I'll Say She Is!* was, in Krutnik's words, a "direct product of vaudeville's entertainment culture," their subsequent vehicles—both on stage and on film—inventively reshaped that culture in line with the sensibilities of New York litterateurs.[77] Kaufman's famous quip backstage at a performance of *The Cocoanuts*—"I may be wrong, but I think I just heard one of my original lines"—may have expressed exasperation at the brothers' inveterate ad-libbing, but there is little doubt among Marx Brothers biographers that his scripts provided a more polished frame for their shenanigans as well as superior wordplay and absurdist puns.[78] Perelman, meanwhile, recalled Groucho's admiration for his use of language. "I knew that [Groucho] liked my work for the printed page, my preoccupation with clichés, baroque language, and the elegant variation."[79] Critical responses to these collaborative endeavors accordingly registered a notable change. No longer lionized as the bricoleurs of past tradition, the Marxes were now roundly embraced in the idioms of comedy's lunatic modernity. Of *Animal Crackers,* critic John Anderson of the *New York Evening Post* described how the team "parade their giddy hallucinations all over the stage, and make their show the most cockeyed of worlds, a comfortable padded cell wherein playgoers may roll about in happy delirium."[80] Robert Benchley celebrated Groucho's punning as the product of a "magnificently disordered mind which has come into its own," while Gilbert Seldes echoed the language of his earlier cuckoo essay in describing the brothers' "totally irresponsible" barrage of "absurdities, lunacies, quips, puns."[81]

Nor were the Marxes unique in following this trajectory. A 1924 *Time* article placed the brothers among a much larger cohort of former vaudevillians who were finding favor among "erudite commentators" for their leading roles in Broadway revues, such as Fanny Brice, Eddie Cantor, Bobby Clark and Paul McCullough, Joe Cook, Leon Errol, W. C. Fields, and Charlotte Greenwood—to name only those listed who would go on to be recruited to talking pictures.[82] Of note here would be Joe Cook, arguably the most acclaimed solo vaudeville comedian of all time, whose career evolved in a kind of one-man parallel with the Marxes'. An extraordinarily versatile performer, Cook first came to vaudeville fame in the mid-1910s in a twelve-minute solo skit titled "The Whole Show" that was possibly an inspiration for Buster Keaton's *The Playhouse* (October 1921): a compressed burlesque of an entire vaudeville show, the bit involved Cook performing every act on the bill himself, trapeze, magic, wire walking, and so on, even to the point of interacting with himself as a song and dance team—in other words, precisely the kind of reflexive pastiche of variety tradition that first drew critics' attention to the Marx Brothers.[83] By the close of the 1920s, he was established as the nation's preeminent "voice of lunacy," celebrated for his famous "Why I Will Not Imitate Four Hawaiians" nonsense shtick (in which he explained why, though adept at imitating two Hawaiians, he drew the line at any more),[84] and a Broadway star of musical comedies like *Rain or Shine* (1928) and *Fine and Dandy* (1930, scripted by Donald Ogden Stewart). The parallel in fact only comes apart with Cook's transition into feature film roles, which he did not sustain beyond a 1930 Frank Capra–directed adaptation of *Rain or Shine*. But in failing to establish a durable presence in early sound features, Cook would prove more the exception than the rule among his Broadway peers, as the next section explores.

"I COVET THE WATERMELON": CUCKOO HUMOR AND EARLY SOUND CINEMA

The migration of New York's cuckoo generation of the 1920s to mass media like talking pictures dramatically raised the stakes for these clowns and litterateurs. What the pages of the *New Yorker* and the Broadway stage could do for the privileged few, the media could do for the masses: cuckoo humor would now conscript the entire nation for its audience. Radio was one pathway to the nationwide spotlight: several Broadway veterans became early radio stars, often as emcees of variety programs that mixed music, comedy, and advertising—Eddie Cantor for the *Chase and Sanborn Hour* (starting 1931), Jack Benny in the *Canada Dry Program* (1932), George Burns and Gracie Allen for the *Robert Burns Panatella Program* (1932), Ed Wynn for the *Texaco Fire-Chief Program* (1932), and many others. Alexander Woollcott became the "Town Crier" for CBS in 1929, while F.P.A. took up duties as a regular panelist on the popular talk show *Information, Please* in 1938. But it was in film that these performers' expanding media reach marked a true turning point in the history of American comedy. Commercial network radio was, after all, a relatively new medium (dating to the 1926 launching of NBC Red): it had no pre-existing tradition of comedy and comedians. In film, by contrast, the arrival of the cuckoo clowns was immediately recognized as a changing of the guard: Hollywood invested heavily in the idea that the innovation of sound would allow it to bring Broadway-style entertainment to the moviegoing masses. In a 1930 essay, "Speaking of Talking Pictures," stage producer Edgar Selwyn complained how "motion picture moguls have combed the stage of its good players, have baited the best with juicy contracts and have left the Broadway mart as dry of excellent actors as a glass eye."[85] Film scholar Henry Jenkins's essential 1992 study, *What Made Pistachio Nuts? Early Sound Comedy and the Vaudeville Aesthetic,* gives a sense of the scale of change:

> In one week, Metro offered contracts to no less than seven cast members of Joe Cook's musical comedy, *Rain or Shine.* Cook hired Ted Healy and His Stooges to replace some of the departing cast members, only to lose them to Hollywood as well. . . . By 1929, Fox alone had more than two hundred stage-trained people under contract; the full scale of recruitment would be hard to estimate. . . . Hollywood [also] bought the rights for many of the period's most popular musical comedies, including *Sally* (with Marilyn Miller and Leon Errol), *Poppy* (with W. C. Fields), *The Cocoanuts* (with the Marx Brothers), *Rio Rita* (with Bert Wheeler and Robert Woolsey), *So Long Letty* (with Charlotte Greenwood), *Whoopee* (with Eddie Cantor), *Simple Simon* (with Ed Wynn), and *Rain or Shine* (with Joe Cook). . . . The hope was that these stars and their stage successes

would represent presold commodities, already familiar to the urban audiences who were the dominant market for the early talkies.[86]

The early sound period has thus come to mean two things in the historiography of film comedy: the displacement of the genteel, sentimental mode favored by the major 1920s slapstick clowns and the corollary advent of a new style of crazy or "anarchistic" comedy derived from the Broadwayites' stage repertoires. In fact, both aspects remain inadequately contextualized. For one thing, the argument that sound-era comedy superseded a gentrified slapstick tradition overlooks the degree to which this displacement had _already taken place_ within metropolitan cultural hierarchies of the 1920s. From the perspective of the metropole, cinema's preeminent clowns were already denigrated as the passé others against which the cuckoo modernity of the new cohort of literary wits and Broadway comedians was secured. Changes in the ranks of Hollywood comedians simply concretized at the level of mass media a change in critical priorities that had already happened. A second observation concerns the historiographic adequacy of the aesthetic categories used to explain the anarchistic mode, although here my claims are necessarily prefatory to the close reading that presently follows. It is Jenkins's influential contention, for instance, that the anarchistic mode of Wheeler and Woolsey, the Marx Brothers, and others was shaped by the "vaudeville aesthetic" that these Broadway performers brought with them to their screen vehicles. For Jenkins, the vaudeville aesthetic connotes a performer-centered mode of theatrical representation whose emphasis on show-stopping virtuosity and display was at loggerheads with the more verisimilar, illusionist aesthetic of classical film narrative: an "aesthetic based on heterogeneity, affective immediacy, and performance confronted one that had long placed primary emphasis upon causality and consistency, closure and cohesiveness."[87] For all its productiveness, however, the notion of a vaudeville aesthetic flattens a near fifty-year theatrical tradition of stage variety into a general, undifferentiated category and in consequence fails to explain the very difference it is called on to illuminate. Why, for instance, did vaudeville's legacy imprint itself in such radically different ways on, say, the silent-era comedies of a Buster Keaton versus the sound features of the Marxes? A historical poetics of early sound comedy should rather attend to the particular inflections that gave comedians and humorists of the 1920s their distinctive flavor and that they subsequently brought to film. If the first step in our project has been to locate those inflections within the specific terms of the era's cuckoo vogue, the second will be to comprehend early sound comedy _in those terms_ and not simply as an epiphenomenon of the catchall category "vaudeville."

I now, therefore, want to continue this investigation by focusing on a double act, Clark and McCullough, that was at the very forefront of the New York theater's cuckoo ranks in the 1920s and subsequently signed to appear in short-subject comedies, first at Fox (1928–1930), then at RKO (1930–1935) (fig. 2). Little remembered today, Bobby

Clark and Paul McCullough in fact rivaled the Marxes in popularity on Broadway in the 1920s and early 1930s, when they starred in major "book" shows like *The Ramblers* (1926) and, their biggest hit, the 1930 revival of George Gershwin's *Strike Up the Band*. In many ways, Clark and McCullough's careers indeed mirror those of the Marx Brothers, and not just because of parallels in their costuming (Bobby Clark painted eyeglasses on his face, just as Groucho painted a fake mustache). Both came up through the tiers of variety entertainment to land starring roles in revues during the 1920s; both won the imprimatur of the peer society's acclaim for their absurdist reworking of low comedy tradition; both were snapped up by major studios—Fox and Paramount, respectively—eager to capitalize on their metropolitan cachet. Yet, where the Marxes' features preserved continuity with the team's stage repertoire—by direct adaptation of their theatrical successes and continuing collaborations with New York litterateurs—Clark and McCullough had a far bumpier road at Fox, which failed to secure adaptation rights for their Broadway shows and consigned them to short subjects with writers who, Clark later complained, were a poor fit for the team's unique style of absurdist incoherence.[88]

Originally from Springfield, Ohio, Clark and McCullough had entered show business in their teens, not as comedians but as acrobatic tumblers in minstrel shows and circus. Their physical training nonetheless made them well suited to vaudeville comedy, which they began to assay in the early 1910s, first as a tramp dumb act, then as a verbal two-act featuring ludicrous animal impersonations. As a four-part profile of Clark, published in the *New Yorker* in the summer of 1947, explained:

> Clark had taken to doing imitations. Most of these concerned little-known animals from the south of some country or other. "An antediluvian oyster, from the south of Bolivia!" he would shout. "Call your special attention to the rigidity of its muscles." Then he would fall down in a limp heap. Springing up, he would elaborate on the species, crying, "A *wild* antediluvian southern Bolivian oyster calling to its *mother!*," after which he would voice a series of horrendous caws, gargles, brays, and moose calls. Clark had added cigars to his props, both on and off duty, and he developed a trick of shifting one around in his mouth to punctuate sentences. "Now, an amphibious creature," he would announce, "found only in the southern Malay Peninsula and the surrounding arch-" (flipping the cigar across his mouth) "pelagoes."[89]

With their move to burlesque in 1917, the team soon developed the personas and stage presences that they would subsequently carry through "unchanged" (according to the *New Yorker*) to Broadway when they achieved breakout success in Sam Harris and Irving Berlin's *Music Box Revue* of 1922: Clark was the duo's lead, a "terrier" of a comic performer, who would develop frenetic bits of prop comedy out of his ever-present cigar and cane, while McCullough's role as stooge entailed little more than following his partner around laughing.[90] The *New Yorker* profile described Clark's characteristic stage entrances: "He sprints on from backstage, brandishing his cane and crouching low, . . . then he rushes to the footlights and

FIGURE 2. Bobby Clark and Paul McCullough. Photo by White Studio © Billy Rose Theatre Division, The New York Public Library for the Performing Arts.

spits his cigar in the general direction of the audience. He catches it, replaces it in his mouth, and spits it out several more times. As a rule, he misses it once, then, with his cane, takes several vicious cuts at the nearby members of the cast."[91] The duo also updated their vaudeville tramp personas, doffing moth-eaten racing coats to become what McCullough later described as "shabby genteel": "Now a tramp has no dignity," McCullough explained, "and false dignity is one of the best comic themes. So instead of playing two down-and-outs, we shifted into playing two fellows on the way down, but still putting up a bluff."[92] Their passage through the Broadway scene of the 1920s was subsequently capped with the record-breaking book show, *The Ramblers,* at the Lyric Theater, for which Bobby Clark (still "smell[ing] of the tan-bark arena," according to the *Times*) received critical acclaim as "one of the most adept stage humorists I ever saw" and the "funniest [comedian] in New York."[93] "He makes low comedy high art" was the opinion of one critic, who thereby confirmed in Clark's nonsense a prestige that the film industry would next try to seize for itself.[94]

That the duo was first consigned to short subjects at Fox, rather than features, reflected no backsliding on that cachet. During the critical years of the transition to sound, shorts were no second-tier support, but rather vital test balloons in the film industry's endeavor to import metropolitan entertainment styles to film (as will be further discussed in the next chapter). Shorts insulated stage comedians and literary wits from the pressures of narrativity that a feature-length film would have imposed, and thereby preserved a space for performances of cuckoo routines that were, in any case, best suited to modular form; in brief, shorts became a first port-of-call in the remediation of metropolitan humor as mass media content. This is evident from the large number of absurdist monologues directly transposed from the stage into short subjects, such as Robert Benchley's *The Treasurer's Report* (ca. May 1928, also discussed in the next chapter), Joe Cook's *At the Ball Game* (ca. July 1928), and Jay C. Flippen's *The Ham What Am* (ca. July 1928), as well as in the verbal cross-talk of double acts like Al Shaw and Sam Lee (*Shaw and Lee in "The Beau Brummels,"* ca. September 1928; *Going Places,* June 1930) and George Burns and Gracie Allen (in *Burns and Allen in "Lamb Chops,"* ca. October 1929), whose wordplay turned on puns, riddles, false syllogisms, and syntactic confusions. ("When were you were born and if so, for instance, if?" asks Shaw in *The Beau Brummels.* "Between nine and ten," Lee responds. Shaw: "That's too many for one bed.") But shorts also became testing grounds for comedians to explore the affordances that the new medium of sound cinema offered for the filmic extension and extrapolation of cuckoo principles.[95] A luminous example is provided at the start of Burns and Allen's *Lamb Chops,* in which cuckoo conventions of verbal disordering are matched with a disorienting approach to cinematic space: the duo search around a living room set on the conceit that they are looking for their "missing" audience, whom they eventually "discover" by looking and gesturing directly to the camera. "There they are, right there, that's them. Say hello to everybody."[96] In a cinematic analogue to the team's comedic wordplay, the materiality of the cinematic signifier (here, a sound stage) supervenes upon and derails the sense (a living room) that it would ordinarily convey.

It is, then, a loss to the history I am tracing that so little of Clark and McCullough's work at Fox survives. Of the fourteen shorts they completed between

1928 and 1930, including some of three and four reels, only one, *The Belle of Samoa* (ca. February 1929), still circulates, and even this is largely built around footage from an extravagant musical number that had been deleted from *The William Fox Movietone Novelties of 1929*.[97] (Another short, *Waltzing Around*—released in early 1929—is held by a private collector.) What evidence there is nonetheless suggests that Fox initially handled the Clark and McCullough series in a manner befitting the duo's prestige, assigning them experienced comedy directors (Norman Taurog, Paul Parrott, and Harry Sweet) and spending lavishly on the films, "with none said to have cost less than $50,000" (over three times what Vitaphone was spending on its shorts).[98] It is also clear, moreover, that the films were, in the main, a series of reproductions of routines they had perfected on the stage, both big-time and burlesque. "Many of their movies," explained the *New Yorker* in its retrospective of Clark's career, "were adaptations of skits they had done in vaudeville, burlesque, or musicals."[99] Sometimes these borrowings were quite direct: the team's first two-reeler, *The Bath Between* (February 1929), was a straight adaptation of a hotel-room sketch the comics had first performed in the *Music Box Revue* back in 1922 (the only surviving script material for this film is tellingly labeled "Their Original Vaudeville Skit"), while other shorts directly lifted jokes from their stage successes (as when *The Diplomats* [February 1929] included some verbal byplay from *The Ramblers*: "How much is a minute steak?" "A minute steak, sir, will cost you five dollars." "Then bring up a couple of split-second sandwiches.")[100] More commonly, the team worked variations on routines they knew from the burlesque playbook: *Waltzing Around* provides an absurdist spin on a burlesque boxing skit; the first part of *Belle of Samoa* is structured as a "con game" routine in which Clark tricks a harem guard into a deceptive wager; and *Beneath the Law* (ca. February 1929) plays out as what in burlesque was called "Irish Justice," a generic term for the courtroom sketches that had been popular since minstrel days.[101] As the *New Yorker* profile explained, "[*Beneath the Law*], Clark believes, is the closest approach, thus far, to transferring burlesque successfully to the screen":

> In it, Clark was the judge who had been arrested for immorality. Every time an attorney opened his mouth, Clark smacked him with a bladder and yelled, "You're trying to inject hokum into this case!" A knock-down, drag-out fight at length developed, involving not only the attorneys and the judge but the jury and the spectators; it was stopped by a suggestion that the stripper do her act for the judge in chambers. Then the judge and the girl made their exit. The short ended when Clark, after a suitable wait, opened his office door and cried, "Case settled out of court!"[102]

These are suggestive hints, to be sure, but there is little enough to suggest anything more than the mimesis of a purely theatrical experience with which early sound shorts have so often been charged. The real achievement of the team's film career would await their subsequent tenure at RKO—after their successful Broadway return for *Strike Up the Band*—when they now had to develop original material for their shorts (most often in collaboration with writers Ben Holmes and Johnnie Grey). Rather than simply submitting older stage routines to the logic of an adaptation, as at Fox, Clark and his team now set about revising two-reel comedic convention in the disorderly image of the duo's lunatic stock-in-trade. They achieved this through a formal approach that was more in keeping with the workings of a kind of absurdist game than with the emplotments of comedic narrative. Scene structure in the best of the RKO shorts often invokes a form of play governed by entirely extemporized rules, as though in pursuit of strange new forms of comedic sport. *Jitters the Butler* (December 1932), for example, has a scene in which the boys are conversing with one another while, for no reason, running counterclockwise around a table. A butler enters the room, briefly joins them for a couple of orbits, and then brings them to a halt. "You can't do this. You can't run round the room this way." "Alright," answers Clark, "We'll run this way," and the trio recommences the game, this time clockwise. Unlike normal games, which are governed by a set of preexisting rules, the game form that Clark and his writers developed at RKO tended toward what the philosopher Gilles Deleuze once described as an "ideal" game, for which "there are no preexisting rules, each move invents its own rules; it bears upon its own rule"; and it did so in a way that, at its extremes, would steer the diegesis toward the collapse of any representational verisimilitude.[103] In a scene from *Love and Hisses* (June 1934), a young man, Harry Knott, is in an office with two private detectives—Clark and McCullough—whom he has hired to stage the abduction of his betrothed, Bunny Bender, so that the two lovers may elope. Bunny telephones to go over the details of the plan and Harry reassures her of the detectives' proficiency. The game that ensues is simple: every time Harry uses a metaphor to describe the detectives over the phone, the duo literally enact the metaphor in question. "They're right on their toes," Harry explains in reference to the duo's readiness, prompting the detectives to "strike toe-dancers' attitudes, raise themselves in the center of the floor, and pirouette," as described in the final script.[104] "We're chinning it over now" (i.e., talking over the plans)—and the pair "leap in the air, grab the cross arms of the chandeliers and quickly chin themselves on the bar [i.e., perform chin-ups] a couple of times."[105] What is up to this point a silly literalness next begins to disentangle the very fabric of the film's fictive coherence. "They can't miss—they're sure fire," Harry reassures Bunny—at which point the detectives pull out their revolvers and shoot at a portrait of Napoleon. Cut to a close-up of the picture: Napoleon fantastically comes to "life," turning to the boys to ask, "Who said so?" (an effect seemingly achieved by placing an actor behind a cutout in the set wall behind the frame). Finally, the topper: "Oh, she did?" continues Harry on the phone. "Well, I'll bet there's a nigger in the woodpile." Clark's detective gingerly pokes at a pile of logs next to the fireplace, from which there bursts a black child who "rushes madly from the scene."[106] (This racist gag is cut from extant prints, which derive from television release versions prepared by RKO in the 1950s.) The systematic process of the game is to introduce into the

VIDEO 1.Clip from *Love and Hisses* (June 1934).

To watch this video, scan the QR code with your mobile device or visit
DOI: https://doi.org/10.1525/luminos.28.3

film a principle that eventually overrules the coherence of its verisimilitude, here summed up in the impossible materializing of linguistic metaphors (vid. 1).

At a larger, plot-wide scale there is a related predilection for what might be thought of as comedic "puzzles" that elaborate the consequences of a single ridiculous premise, often involving a character governed by an overriding fetish or fixation: a butler motivated solely by the pleasure he takes in being kicked in the pants in *Jitters the Butler,* a pet pig addicted to mints in *In a Pig's Eye* (December 1934), or a judge with a positively erotic yearning to eat watermelon in *Love and Hisses* (working title: "I Covet the Watermelon")—each of which inserts an entirely alien component into ordinary conventions of motivation and psychology. Verisimilitude is here dismantled by the Rube Goldberg–style mechanics of a narrative riddle: How might a pig's mint addiction derail a business deal between an inventor and two phony counts? How might a man's desire to eat watermelon be made to resolve the prohibition on Bunny and Harry's marriage? (The answer? The detectives mislead the judge into declaring his love for Bunny's mother by tricking him into thinking he's talking about a watermelon.) As with Clark and McCullough's "game" approach to comic business, what the RKO team here installed was a system for derailing fictive sense that once again permits of a Deleuzian reading. In one of his late texts, "Bartleby; or, The Formula," the philosopher defined the narrative tactics of Melville's 1853 story as the working out of a "formula," similarly linked to a protagonist's fixation (the scrivener's "I prefer not to") whose elaboration hollows out the representational system of the fiction, of causes and effects, plausible behaviors and their motivations. Bartleby's repeated refusals, Deleuze notes, introduce a kind of blockage or foreign element into the cogs of storytelling convention, the effect of which, like Gregor Samsa's metamorphosis in Kafka's novella, is to submit narrative instead to the working out of a puzzle: "What counts," he writes, "is that things remain enigmatic yet non-arbitrary: in short, a new logic, definitely a logic, but . . . without leading us back to reason."[107] Something of this may be found in the Clark and McCullough short *Snug in the Jug* (November 1933), whose plot seems to have been conceived from the outset to produce this kind of conundrum. The earliest sketches by writer-director Ben Holmes carefully outline the premise of a narrative double bind: Clark and McCullough play recidivist prison inmates who, each time they are released, are advised to steer clear of their nemesis Slug Mullen (played, in the finished film, by Harry Gribbon); each time, however, they immediately encounter him and are returned to jail. Here is Holmes's outline for the film's opening, from a draft dated June 29, 1933:

1. The judge hammering, smiling "thirty days." LAP DISSOLVE.
2. EXTERIOR PRISON ENTRANCE. Warden ushers Clark and McCullough out—short dialogue "Free again—keep away from Slug Mullen—nemesis, etc." Clark and McCullough exit.
3. DESERTED STREET NEAR PRISON. Slug Mullen with pocketbook, meets Clark and McCullough—gives them pocketbook—exits. Clark and McCullough examine contents. Policeman in—drags them out. LAP DISSOLVE.
4. The judge, smiling, raps "sixty days." LAP DISSOLVE.
5. EXTERIOR PRISON DOOR. Warden again ushering Clark and McCullough out—short dialogue "Free again—keep away from Slug"—ten dollar gift. McCullough doesn't want it. Clark chastises him—they take money—exit.
6. EXTERIOR VEGETABLE AND FRUIT STORE. Slug Mullen attempting to force racket on owner. Italian owner objects. Slug throws grapefruit through window. "That's only a sample of what you're going to get." Owner throws vegetables. Clark and McCullough enter—throw vegetables—pumpkin through window. Policeman in—Clark and McCullough arrested—dragged out. LAP DISSOLVE.
7. The judge smiles, raps "ninety days." LAP DISSOLVE.
8. EXTERIOR PRISON. Once more Warden ushers Clark and McCullough out—free again—gives them money—bethinks himself of job—gives them job tacking cards. Clark and McCullough exit.[108]

The boys can only stay out of prison by avoiding Slug Mullen, but it is only in prison that he can be avoided. How, then, to avoid Slug and be free? In the initial outline, Holmes cheats on the dilemma by removing one of its horns: the duo opt out of freedom by finally electing to live in jail (the "only place they can avoid Slug Mullen").[109] Two days later, in a draft continuity dated July 1, Holmes cheats again, this time by dissolving both horns: the film ends with Slug *and* the boys in jail. But it is only with the final script, a week later, that the writer keeps faith with his own riddle. As in the initial outline, the duo escape into prison to avoid Slug only to find, in a twist on the draft continuity, that Slug is *already* there waiting for them. The final script details that twist and its ensuing action as follows:

78. INT. PRISON—DAY

FULL SHOT—The same cell in which Slug Mullen has been locked. Clark and McCullough entering the cell. From the inside, Clark reaches his hand out, locks the door and throws the keys down the hall. Clark beats on the iron bars with his cane.

CLARK

Well, Blodgett—at least we're safe from Slug Mullen.

MAC

You said it.

CUT TO

80. INT. CELL—DAY

CLOSE SHOT—Clark and McCullough standing near the bunks. Slug Mullen raises himself into view from the top bunk, glares down at them.

SLUG

Oh yeah!

A big take from Clark and McCullough as they see Slug.

CUT TO

81. INT. JAIL—OUTSIDE CELL—DAY

Clark and McCullough rush forward toward the CAMERA, apparently running right through the iron bars. They stop in the foreground. Slug leaps down from the bunk after them, rushing up against the iron bars, and nearly knocks himself out. Clark and McCullough laugh.

CUT TO

82. INT. JAIL—OUTSIDE CELL—DAY

CLOSE SHOT—Clark and McCullough as they gaily call to Slug.

CLARK AND MAC

(giving the cross fingers to Slug)

Stone walls do not a prison make

Nor iron bars a cell.

They go gaily, skipping down the hall, as we

FADE OUT.[110]

Clark and McCullough here become possessed of a kind of crazy or magical excess that violates the very spatial alternatives (in jail/out of jail) on whose exclusivity the foregoing action has depended. Which is to say that the conundrum of Slug's inescapability is finally resolved, not within the framework of what is permitted to narrative, but by the materiality of the medium itself; that is, in a special effect that allows the duo to pass through the iron bars of the prison cell (figs. 3 to 5).

FIGURES 3–5. Clark and McCullough pass through the iron bars of a jail cell. Frame enlargements from *Snug in the Jug* (September 1933).

The early sound vehicles of Hollywood's Broadway cadre indeed abound in moments like this "magical" jailbreak, in features as in shorts: a tattoo of a dog on Harpo's chest suddenly barking (by virtue of an animated special effect) in the Paramount feature *Duck Soup* (1933); W. C. Fields on the receiving end of a faceful of fake snow every time he opens the cabin door in the Sennett two-reeler *The Fatal Glass of Beer* (March 1933); a bust coming to life to join the villains' laughter in Wheeler and Woolsey's *Diplomaniacs* (1933). Whereas conventional classical filmmaking sought to make textual signifiers cohere around the illusion of a fictive reality, early sound comedy evinced what Henry Jenkins has called an "expressive anarchy" that played with those very signifiers as a source of amusement.[111] But in light of the foregoing, it seems clear that this anarchy is to be understood not simply in the effort to bring a generalized "vaudeville aesthetic" to film—Jenkins's position—but rather in the endeavor to remediate the more specifically absurdist forms of the cuckoo vogue in filmic terms: expressive incoherence on the scale found in these films belongs less to vaudeville per se than to the lunatic inflection of 1920s humor that turned toyingly to the possibilities of its own media of expression, dismantling signifying processes into pure form.[112] The paradox, next to be considered, is that the film industry had thereby invited a transposition of comedic codes whose implications for filmic ones would not long be tolerated.

"KEEP THEM DOING SLAPSTICK GAGS": THE DEMISE OF CUCKOO HUMOR IN MOTION PICTURES

What does it say about the conditions of early sound-era filmmaking that Clark and McCullough never made the splash that their Broadway reputation promised? There can be little question that their start at Fox was badly mishandled. At a time when the team might easily have graduated to features, Fox failed to secure rights to their stage hit, *The Ramblers,* which was instead picked up by RKO and shot as *The Cuckoos* (1930), starring that studio's then-resident cuckoo team of Wheeler and Woolsey (this being prior to Clark and McCullough's own tenure at the studio). Nor can there be any question that their subsequent stint at RKO failed to smooth the team's relations with the Hollywood studios. As the *New Yorker* profile explained, Clark

> is still badly soured on Hollywood; he has never felt right about the place since the team's last appearance there, early in 1935. Signed by RKO after a spell of Broadway shows, the two men . . . were met at the station by a band and cheering section and carried bodily into the office of the president. "Here they are, Chief!" cried a hysterical press agent. "*Clark and McCullough!*" "Ah, yes," said the executive with a clammy smile of welcome, "the dancers."[113]

Clark's evident resentment notwithstanding, it was not the perception of mistreatment but a more tragic event that ended the team's screen career—and, indeed, the team itself—when McCullough committed suicide in 1936. After a hiatus of several months, Clark would return to the stage as a celebrated solo performer, eventually acquiring a reputation as "the last of the great clowns," but returned to the screen only once, for the *Goldwyn Follies* of 1938.[114]

None of this, however, speaks to the paradox that came to settle over their work in film: that, despite the initial fanfare, the cachet of their screen appearances eventually slipped far below that of their stage reputations. Three-time Clark and McCullough RKO director Sam White addressed precisely this dilemma in a later interview: "The thing about Clark and McCullough was that when you directed them on the set, they were hysterical, especially Bobby. I used to think that the scenes I was making would split my gut. When we got it on film, it wasn't funny. They just never came off funny. They exuded some chemistry in person that never came off on the screen as it should have." The only way he and the unit's other directors (including future musical helmer Mark Sandrich) "could ever make them funny was to keep them doing slapstick gags all the time."[115] Some of the later RKO shorts testify to this shift in approach, the absurdity of the team's expressive incoherence replaced by a more straightforward physical freneticism.[116] A case in point would be the team's final film, *Alibi Bye Bye* (July 1935), which revisits the burlesque "hotel scene" they had already used for *The Bath Between* at Fox and *The Gay Nighties* (June 1933) at RKO, albeit with a switch of emphasis that belies the apparent continuity: whereas the earlier RKO short generated humor from the deliriously implausible escalation of a detective's efforts to apprehend a burglar (culminating in a machine-gun shootout and motorcycle chase through the hotel's corridors), *Alibi Bye Bye* reclaims the team to a more straightforwardly knockabout vector, bodies in motion running from room to room. (The film, notes slapstick historian Steve Massa, "probably set the record for the most doors slammed in any farce film.")[117] The stage versus screen paradox of Clark and McCullough's career is, then, perhaps best framed less in terms of the imponderable of the team's failure to "register" for the camera than in terms of the particular challenge of sustaining innovations in one medium (in Clark and McCullough's case, variety theater) in the different institutional circumstances of another (film): despite their inventiveness in adapting two-reel comedy to their cuckoo stock-in-trade, Clark and McCullough's creative development remained prey to the proclivities of filmmakers who had cut their teeth on the conventions of an earlier era. Sandrich, for instance, had first come up the ranks in the late 1920s as a director of silent two-reelers starring Lupino Lane, one of the British stage's preeminent exponents of music hall-style acrobatic pantomime; Sam White was the youngest of the White brothers, whose careers as slapstick producer-directors—discussed in more detail in chapter 5—dated to the Keystone era. "Making them funny," for filmmakers like these, often meant a return to straight slapstick.

Yet the case of Clark and McCullough would signify nothing more than a single instance of creative tapering if it did not mirror a pattern exemplified across all the cuckoo comedians recruited to Hollywood. Even though Hollywood's transition to sound depended heavily on Broadway humorists for short subjects in its initial stages, as the 1930s progressed, cuckoo reflexivity increasingly gave way to more conventional comedic emplotments. What animation historian Nic Sammond has noted of cartoons during this era held for live-action comedy too: sound film soon came to demand the "spatial enclosure" of narrative space—that is, a return of classical standards of narrative construction that consigned these trickster figures to coherent fictive worlds in which mayhem was now directed inward, within the world of the fiction, rather than outward, at the framework of the fiction's coherence.[118] Within the standard histories, the transition is symbolically marked by the Marxes' passage from Paramount to MGM, from the anarchistic excesses of the finale to *Duck Soup* to the more gentrified romance plots of *A Night at the Opera* (1935) and *A Day at the Races* (1937). But the pattern was first and most vividly drawn in the case of those comedians hired to the maw of live-action shorts—like Clark and McCullough—where the very pace of production more quickly forced a turn to narrativization. The trajectory emerges clearly, for example, in vaudeville comedians George Burns and Gracie Allen, who first appeared in film before the Vitaphone cameras in the late summer of 1929 to perform their celebrated cross-talk act, "Lamb Chops," their stage staple since 1926. Subsequently signed to Paramount, the duo proceeded to make eleven short films between 1930 and 1933, requiring a team of writers to come up with a scripting strategy that could accommodate this dramatically increased pace of production. The approach taken was to locate Burns and Allen within familiar fictive environments— drugstores (*Pulling a Bone*, January 1931), hotels (*100% Service*, August 1931), hospitals (*Oh, My Operation*, January 1932)—which provided a basic framework for their trademark wordplay even as they confined that wordplay to the bounds of narrative verisimilitude. Burns and Allen's characteristically "cuckoo" disregard for cinematic illusion in the earlier shorts—the way they would puncture the enclosure of cinematic space by waving at the camera or calling attention to the sound stage—seems to have simply vanished midway through their Paramount series in favor of a more rigorous diegetic coherence.[119] The trajectory was also exemplified by Robert Benchley in his six films for Fox between 1928 and 1929: his first two—*The Treasurer's Report* and *The Sex Life of the Polyp* (July 1928)— were absurdist monologues delivered direct to camera, the former based on a stage routine he had developed in 1922, the latter developed from a number of early essays ("The Social Life of the Newt," "Do Insects Think?," and "Polyp with a Past"); the remaining four were more familiarly plotted comedies that adapted Benchley's essayist persona to the narrative paradigms of middle-class "situation" comedy (learning to drive, gardening, etc.).[120]

But the example that comes closest to Clark and McCullough, despite great differences in performative style, is that of Ted Healy and his Three Stooges. First cast into the limelight in Broadway revues like J. J. Shubert's *A Night in*

Spain (1927) and *A Night in Venice* (1929)—where they were variously billed as "Ted Healy and His Racketeers" or "Ted Healy and His Gang"—Healy and the Stooges were infamous for running ragged over the fictive coherence of their librettos: Healy would serve as emcee, the "Racketeers" as interlopers who would haphazardly barge into scenes to disrupt Healy's efforts to manage proceedings. (For *A Night in Venice,* for example, the Stooges were first introduced as hecklers planted in the audience whom Healy invited onstage to wrestle a man in a bear suit. The "unsightly fellows spent some 15 long minutes slapping each other's faces . . . following which one of the slappers wrestled with the bear and had his clothes ripped off.")[121] Signed to film—first at Fox for the Rube Goldberg–scripted feature *Soup to Nuts* (1930), then, three years later, at MGM for a series of five two-reel comedies and cameos in feature-length musicals—Healy and his team were initially cast in films that structurally replicated their revue appearances. Showbiz settings provide a framework for nonintegrated song and dance numbers (some performed by frequent MGM costar Bonnie Bonnell, others consisting of recycled footage cut from the studio's features), while the Stooges maintain their stage function as nonsensical agents of unmotivated disruption, showing up out of nowhere to spray Healy repeatedly with water in *The Big Idea* (May 1934) or bursting on stage to yell and shout over Healy's rendition of "Dinah" in the revue short *Plane Nuts* (October 1933).[122] This is slapstick, true, but, as in the team's stage turns, it is a slapstick "gone cuckoo" by virtue of its intrusion into an alien representational context, by the pure mechanism of a random intrusiveness that rebels against any representational sense (fig. 6). And it was precisely this quality that would be sacrificed in the Stooges' move to Columbia Pictures, sans Healy, in 1934, where they commenced a quarter-century run in almost two hundred short subjects, making them the quintessential emblems of slapstick's sound-era legacy. As will be explored in more detail in chapter 5, the path to Columbia had the effect of placing the Stooges in the hands of veteran comedy filmmakers like Del Lord, whose creative enthusiasms lay rather in the throwback direction of Sennett-style roughhousing. As if the jailers had locked up the cells, the Stooges were now shut off in their own slapstick universe, their instinct for nonsensical chaos channeled into conventional slapstick fiction rather than disordering the sense of the fiction from without.

*

A reference to sociologist Pierre Bourdieu brings this chapter home: the introduction of any innovation into a given field, he argues, is never a matter of a "pure confrontation with pure possibles" since it depends on the "system of possibilities" that already defines that field (the interests of the various agents involved in the process, the legacy of techniques and approaches bequeathed by the past, as well as the production processes in which works are caught up).[123] In the field of film comedy, the innovation of sound may be said to have

FIGURE 6. The Stooges appear only three times in *The Big Idea* (May 1934), first marching into Healy's office with trumpets from which they spray him with water, second the same thing with French horns, third with tenor saxophones, with no explanation of who they are or what they are doing.

opened a window for experimentation (for testing "pure possibles") that the industry's practices and proficiencies (its preexisting "system of possibilities") quickly pulled shut again. The media diffusion of cuckoo humor might in fact be said to have met its Waterloo in motion pictures three times over—in its clash with swiftly reestablished standards of narrative enclosure, in the comedic preferences of directors who had learned their trade in the silent era, and especially in shorts, in the rapidity of production processes that exhausted the cuckoo comedians' fund of stage material. The unimpeachable modernity of Broadway clowns and humorists may have rendered the conventions of silent slapstick passé, but it could not dispatch them. Like a dead planet, the comedic templates of an earlier era continued stubbornly on their path, even pulling Broadway challengers like Clark and McCullough and the Marxes into their orbit.

Not that the cuckoo style completely disappeared from film. It staged a couple of last hurrahs in two Universal-produced anarchistic features, both released in 1941: W. C. Fields's *Never Give a Sucker an Even Break* (Fields's last script) and the screen adaptation of Ole Olsen and Chic Johnson's zany Broadway revue, *Hellzapoppin'*. It also sustained a shadow existence as a means of class characterization in the

Depression-era screwball style, for which absurdism became a fictive marker of upper-class lifestyles—whether negatively, as the sign of the elite's irresponsible dissociation from the real world (in *My Man Godfrey,* 1936), or, more positively, as the sign of their repudiation of moribund class rituals in a supposedly democratic spirit of play (e.g., *Holiday* or *You Can't Take It with You,* both 1938). Nor, moreover, would it be fair to single out cinema for blunting the edges of American revue comedy's leading lights. What was true of early sound comedy, for example, was also true of commercial network radio, whose parallel birth (with the 1926 launch of NBC Red) was similarly nurtured by Broadway talent. Similar to their short-subject confrères, radio comedians were forced to come up with representational frameworks designed to generate new comic material for shows broadcast on a weekly or even twice weekly basis. Some succeeded—like Jack Benny, who, in collaboration with writer Harry Conn, developed a framework featuring recurrent characters interacting in changing comic "situations," thereby anticipating the format of the television sitcom.[124] Others failed—like Robert Benchley, again, whose short-lived show *Melody and Madness* (1938–1939) saw him cede script control of his monologues to sponsor-assigned writers whose formulaic gags, one critic averred, "could have been delivered by any radio comedian."[125] The lessons to be drawn from motion pictures are, in this sense, incomplete without acknowledging the broader pattern of cuckoo humor's commercially driven media spread and the changing semiotics of comedic representation that each medium produced. In network radio as in synchronized sound film, the Depression-era mediascape was fueled by the appropriation of leading metropolitan comedians who greased the wheels of new and evolving media platforms, even as their distinctive personas were thereby reshaped and sometimes dissipated.

What remains to be underscored, however, is the unusually prominent role that shorts played within these processes. It was in the field of live-action comedy shorts, I have suggested, that the mass mediation of the cuckoo style was first auditioned; in shorts, too, that the endeavor first comprehensively broke down. In that missed connection was already foretold the field's future identity: no longer a fertile terrain for the leading edge of comedic innovation but, we will see, a home for veteran slapstick filmmakers to grow old together; put another way, a refuge from the very modernity that cuckoo humor had once described. The next piece in the puzzle requires that we turn to the short-subject sector as a whole to consider how it took shape from changing industry strategy during Hollywood's rough ride through the Depression.

"The Stigma of Slapstick"

The Short-Subject Industry and Its Imagined Public

In June 1929, the Vitaphone Corporation produced a remarkable short, *Don't Get Nervous*, which cannot but strike a viewer for its reflexive engagement with the problems of representation and address confronted by so many early sound shorts. Like other Vitaphone reels of the time, it consists of a vaudeville performance—here, a solo or one-act by comedian Georgie Price—staged frontally before the Vitaphone cameras. Yet what makes for distinctiveness is the way its opening varies the standard format to offer a seeming "behind-the-scenes" glimpse at the short's production. *Don't Get Nervous* begins with a couple of shots showing the crew preparing Vitaphone's Brooklyn soundstage for filming. Next, in the third shot, Price strides in and, in a state of agitation, demands to see "Mr. Foy" (ex-vaudevillian Bryan Foy, the actual Vitaphone unit supervisor), who is in turn ushered in and, in medium two-shot, asks Price what's upsetting him (fig. 7). "What's upsetting me?" Price responds, and he lists the problems:

> This studio. Thousands of fellows running around. All this excitement. Hanging lights. Hanging microphones. Folks fixing things around here. You know it's different in the theater. In a theater there's a wonderful audience. As soon as I walk out on the stage, I start to sing. I can look at their faces and tell whether they're with me or not. But here it's different. There's no atmosphere. No audience. No nothing. I'm nervous about it, Brynie [Foy's nickname].

singing

"Oh, aren't you foolish," Foy replies, breaking the fourth wall to point toward the camera. "Why, you've got a real audience right here." Does he mean us, the geographically and temporally dispersed movie audience? Apparently not, for the film now cuts in a reverse shot to present a technological apparatus: the Vitaphone camera booth, two operators peering out, upon which is affixed the sign "THIS IS YOUR AUDIENCE" (fig. 8) The framing of the booth, which fills the image, creates a starkly

FIGURES 7–9. Bryan Foy (left in the first shot) provides an onset audience to calm Georgie Price's concerns about performing for the Vitaphone camera. Frame enlargements from *Don't Get Nervous* (ca. July 1929).

VIDEO 2. Clip from *Don't Get Nervous*.

To watch this video, scan the QR code with your mobile device or visit DOI: https://doi.org/10.1525/luminos.28.4

reduced sense of depth that underlines the abstract impersonality by which Price is so clearly discomfited. Back to the two-shot: "You call that an audience?" the comedian complains. The penny drops for Foy. "Oh, I see. You want a real flesh-and-blood audience," and he invites the crew to assemble on one side of the Vitaphone camera to serve as an on-set "flesh-and-blood" harbinger for Price's ultimate audience in movie theaters (fig. 9). His concerns allayed, Price begins his act for the assembled crew, and the short proceeds in standard fashion—only now with the understanding that Price is performing *not* for the camera but for the assembled technicians and stagehands, who have become proxies for the film spectator (vid. 2).

There are issues of no small critical interest here, foremost among which is surely Price and Foy's endeavor to reproduce the "liveness" of the vaudeville stage within the time and space of a filmic performance. "Canned" vaudeville such as Vitaphone's imposed a total separation of the time-space of an originating performance from the time-space of its reception, and with that, the withholding of the "atmosphere" of copresence upon which vaudevillians thrived.[1] But *Don't Get Nervous* attempts to manage those concerns by overtly altering the framework within which Price's routine occurs. The camera as the impassive tool of a merely mechanical inscription is displaced by the stagehands, whose presence thus folds the moment of performance and its reception into a single space (the Brooklyn studio); the spectator of the film is, consequently, asked to view the performance not as a technologically mediated separation, but instead by identifying imaginatively with the surrogate "live" audience that has been implanted within the space of the film's production. What is explored here are not only the performance conditions necessary to alleviate Price's discomfiture, but also the conventions of filmic representation needed to acclimate viewers of early talkie shorts to vaudeville-style presentations in which, as film historian Charles Wolfe notes, the "relationship between actor and the audience [was] by necessity imaginary."[2]

Although these themes will emerge tangentially in the following pages, they are not the paths that will primarily be pursued in this chapter. For I would like to engage Price's perplexity about his audience in a way that is at once more material in social terms and more metaphorical in terms of the film industry's broader insecurities about *its* public: Who was the imagined audience for sound cinema after all? And how were industry assumptions about the moviegoing public manifest in the production and marketing strategies adopted by short-subject producers? From this perspective, the confounding slogan confronted by Price below the camera booth's window—"THIS IS YOUR AUDIENCE"—and his uncertainty about whom he should be performing for become signifiers of a broader impenetrability that baffled the short-subject industry as it sought to eke out a role for itself within changing film industry practices both before and after the upheavals of the Depression.

Any cultural industry will of course operate with a certain idea of the public for its product, but there are two vectors through which that idea may be approached. There is, first, the empirical "who" of the public, the actual audience for the industry's products (women/men, young/old, etc.), as this might be determined by, say, a statistical analysis. But there is also an "imagined" public or, perhaps better, an "idea" of the public as rhetorically framed by that industry, as hailed by a particular address. What I have in mind here is the way any culture industry will communicate with its imagined public in terms of projected social values and intentions—for instance, by addressing itself to consumer fantasies, nationalist fears, civic ideals, and so on. Such a public, as Michael Warner has argued, is best thought of not as a statistical entity but as a "space of discourse" that "exists only . . . *by virtue of being addressed*," or as a "social space created by the . . . circulation of discourse."[3] In the matter of the film industry following the coming of sound, there are in this respect two main points. First, the Hollywood industry had by this point long understood the empirical "who" of its audience in terms of the rather nebulous concept of the "masses"—a socially and ethnically heterogeneous white audience whose imaginary cohesion rested on the structural exclusion of nonwhite Americans.[4] This assumption of a mass audience was well established within the industry's self-idealization as a democratic art and would be further enshrined during the 1930s, in studies like Margaret Thorp's *America at the Movies* (1939), which posited sound cinema as a new form of shared symbolism, spanning differences of class, generation, and region (but still not race).[5] The second point: there was a clear transformation, before and after the impact of the Depression, in the nature of the studios' *address* to that mass public, a shift in the ideation of the "masses" as the object of Hollywood's discursive rhetoric. As film historian Catherine Jurca has argued, Hollywood's marketing strategies of the latter 1930s witnessed a concerted effort on the film industry's part to reconceive its relation to its public: in the earliest years of sound, the Hollywood studios had characteristically addressed itself to a public envisioned in terms of hierarchical separation across urban/regional lines, but around the mid-1930s, this hierarchical model came to be replaced by a more equivalential one, which conceived the moviegoing public in civic terms, united in the era's populist imaginary.[6]

These two modes of address implied very different constructions of the mass audience—the former inflected by Jazz Age hierarchies of taste and the top-down dissemination of metropolitan-style entertainment, the latter by New Deal–era ideals of civic spectatorship and the construction of shared popular identities. Both, further, bookended the operations of Hollywood's short-subject sector during this period, where they were enshrined in the competing market strategies adopted, first, by Warner Bros. for the launching of its Vitaphone sound shorts and, second, a decade later, by MGM's revamped short-subject unit under Jack Chertok. In this chapter, accordingly, I want to use Hollywood's changing conceptions of its public as a thematic for tracking the broad contours of the short-subject industry's development during these years and the position of the slapstick short within them. The goal is to offer a historical understanding of the short-subject industry as structurally connected to Hollywood's evolving imaginaries of the "masses" and the varying patterns of social difference and distinction that they sought—or, in the case of slapstick, now failed—to regulate and manage.

"SOUND CAME ALONG AND OUT WENT THE PIES": VITAPHONE'S BROADWAY STRATEGY AND THE DELEGITIMIZING OF SLAPSTICK

Writing in March 1930, the *Exhibitors Herald-World's* Broadway columnist, Peter Vischer, offered a witty description of developments in short-subject comedies since the coming of sound. "Sound came along and out went the pies," Vischer began.

> No longer was it possible for the average American male mind, aged 14, to enjoy the spectacle of features emerging from a gouey [*sic*] crust or to project himself, figuratively, into the person who was giving the other person, usually a Mr. Milktoast, a lusty boot in the slats. No sound had come in and the day of the [vaudeville] act arrived.
>
> Mr. Picture-Goer, for his comedy entertainment, had to watch a vaudeville actor play the banjo and sing songs that should have been burned years ago. . . . Then came another change. Producers woke up to the fact that acts were not exactly hot; that what might be considered the novelty of sound was no excuse for bum vaudeville. They began to put into the production of their short subjects the same happy robustness that marked them before the microphone reared its trembling magnet before the stuttering player.
>
> Pies actually came into use again. Now you can hear them plop, as well as see them squash. Not that pies are prevalent today; but the spirit that prompted them is. Short comedies have retrieved their schoolboy virility. They are alive, brusquely humorous and often broad. They are productive of belly laughs rather than wan smiles. And that's what they should be.[7]

"Sound came along and out went the pies": Vischer's thumbnail sketch productively recasts the challenge confronted by slapstick filmmakers of the early talkie era. What in retrospect has appeared to later critics as a linear teleology of decline is presented here as a restructuring of what we have called, following sociologist Pierre Bourdieu, the "field" of film comedy production. The concept of a field is in fact directly relevant to the present analysis; for, as developed by Bourdieu, it encourages consideration of how any sphere of cultural endeavor comprises a structure of individuals and groups—for instance, filmmakers and studios—"placed in a situation of competition for legitimacy."[8] Reading Vischer through Bourdieu, the innovation of sound can thus be seen as introducing a new axis across which struggles over legitimacy were waged, bringing about the "day of the vaudeville act" in talkie shorts that, at least temporarily, threatened to dethrone slapstick as the dominant format of comedy shorts. It is, accordingly, on the terrain of legitimacy—of cultural value and worth—that analysis of sound's impact on the short-subject sector will begin: How was the technological innovation of sound yoked to the question of cultural value? And how was this manifest in the "day of the vaudeville act"?

As it was in shorts that sound was initially introduced, these are questions that lead in the first instance to the studio that most successfully spearheaded the talkie revolution: Warner Bros. With its first program of sound-on-disc Vitaphone shorts accompanying *Don Juan* on August 6, 1926, Warners had initially sought to impress by appealing to traditional highbrow standards: the overture from Wagner's *Tännhauser*, performed by the New York Philharmonic; tenor Giovanni Martinelli's aria from *I Pagliacci*; sopranos Marion Talley and Anna Case performing music by Wagner, Dvořák, and Beethoven—with only Roy Smeck's solo on the Hawaiian guitar offering lighter musical fare. The model here was to frame technological innovation as a source of cultural dissemination, as became explicit in the evening's opening short, in which Will Hays, president of the Motion Picture Producers and Distributors of America (MPPDA), described sound cinema's promise. Sound film, Hays stiffly asserted, would exercise "an immeasurable influence as a living, breathing thing on the ideas and ideals, customs and costumes, the hopes and the ambitions of countless men, women, and children."[9] Extending a rhetoric of technologically enabled uplift that already informed radio broadcasting, Hays asserted that "the motion picture is a most potent factor in the development of a national appreciation of good music" and defined sound cinema, like radio, as a medium capable of transcending geographic dispersal: "Now that service will be extended as the Vitaphone shall carry symphony orchestrations to the town halls of the hamlets."[10] Hays's words were fully in line with Warner Bros.'s promotional discourse, which elsewhere described the Vitaphone as a force of cultural democratization that would make "available to audiences in every corner of the world the music of the greatest symphony orchestras."[11] Nor was this just rhetoric. No other studio invested as heavily in opera during the initial conversion period as Warners, which boasted an exclusive contract with the Metropolitan Opera House granting rights "to engage any of [its] singers and musicians."[12] Between 1926 and 1932, Warners produced a total of some sixty-five opera shorts— most of which were completed by the end of 1927 and held for later release—while other studios produced just a handful of similar films.

Yet despite these initial highbrow endeavors, signs of variation in Warners' strategy were manifest as early as the second program of Vitaphone shorts (October 7)— which showcased more popular, comedy and jazz-oriented routines by Al Jolson, George Jessel, the double act of Willie and Eugene Howard, and songstress Elsie Janis, foretelling the shift to a policy of Broadway-style variety that would soon come to dominate Vitaphone's output. Music from the leading big band orchestras; monologues and two-acts by big-time revue stars and vaudevillians; comic and dramatic playlets; and a wide assortment of novelty performers, such as the five-year-old torch singer Baby Rose Marie and Sol Violinsky's simultaneous playing of the violin and piano—these would become the preferred performance types, and for the 1928–1929 season, Vitaphone refurbished its Brooklyn production stages (the old Vitagraph studios) the better to tap the Broadway talent pool. By 1929, publicity for Vitaphone was drastically minimizing sound technology's initial association with opera and classical music to foreground forms of entertainment in line with the rhythms of big city life, even retitling its series *Vitaphone Varieties* to that end. As Charles Wolfe has noted, "The decision to retitle the series 'Vitaphone Varieties' . . . formally acknowledged" the change in strategies, marking a shift toward an aesthetic "derived from vaudeville, with a premium placed on the diversity and novelty of ten-minute acts grouped together in various clusters."[13]

The importing of the era's leading "cuckoo" stage comedians, examined in the previous chapter, really begins with this "Broadway strategy" on Vitaphone's part, and our discussion there consequently sheds light on the cultural identity that was thereby being claimed for sound cinema. It would be a mistake, for instance, to assume that Vitaphone's shift from opera and classical music to a Broadway model can be characterized as a shift across the axis from "high art" to "popular" standards, since the very notion of Broadway-style entertainment was a symptom of the displacement of those very distinctions. As outlined in the previous chapter, New York's Jazz Age ethos of cultural rejuvenation encouraged an alternative interpretation of distinction than that through which the genteel classes had formerly sought to police the boundaries of culture. Rather, the Broadway revues and nightclubs of the 1920s were testing grounds for a newly secular entertainment culture wherein Victorian ideals of restraint and self-discipline ceded to an insouciant and expressive metropolitanism.[14] Similarly, performers captured by the Vitaphone belonged not to the realm of what had once been working-class variety or cheap vaudeville, but to the showier firmament of musical revues that had emerged from the pioneering efforts of New York nightlife entrepreneurs like Florenz Ziegfeld and Jesse Lasky to repackage variety for the urbane "peer society." The emphasis in the marketing of the Vitaphone shorts was thus, by around 1928–1929, firmly on notions of urbane exclusivity and distinction ("Vitaphone links your theater to Broadway . . . Broadway—Mecca of millions now round the corner resort of all America, thanks to Vitaphone!"), establishing a strategy of appeal that spread swiftly throughout Hollywood during the early sound era.[15]

As a publicity tactic, this "Broadway strategy" provides further evidence of New York's ascendancy as the nation's barometer of, in F. Scott Fitzgerald's turn of phrase, "what was fashionable and what was fun."[16] But it also reaffirms the shift toward a new paradigm of cultural hierarchy in which class and ethnic difference were increasingly overlaid by sectional associations, enshrined in the emergent division separating metropolitan sophistication from small-town hokum. What further deserves to be stressed, however, is how these developments thereby altered the rhetoric of cultural uplift in relation to the uses of mass media: whereas late Progressive Era reformers had advocated what might be thought of as a *trickle-down* model of cultural dissemination, which used mass media to bring "high" culture to the "masses," Jazz Age discourses favored a *broadcast* model, which conceived of mass media—first radio, then cinema—as a means to close the cultural "gap" created by distance from metropolitan centers.[17] It is thus significant that an earlier class-based language of uplift that, during the single-reel era, had sought to transform the nickelodeon into, for example, "The Poor Man's Elementary Course in the Drama," was now framed in *spatial* terms that celebrated Vitaphone as a means of transcending the "miles that used to separate you from the Street of Streets."[18] As one ad from this period announced, "Broadway has burst Manhattan's boundaries. . . . No longer must you travel to New York to see the greatest stage attractions. Just—Step around the corner . . . and you're on Broadway!" (fig. 10).[19]

The emerging Broadway strategy was also likely a strategic response to changing market conditions. Before 1927, the majors could afford to invest in a variety of productions suited to a range of tastes, but, as Paul Seale has argued, a sudden decrease in profits in 1927 led to an industry-wide downscaling and concentration on surefire profitability.[20] It is evident, for example, that the turn away from conventionally "highbrow" fare resulted in broader success for the Vitaphone reels, at least to judge from one early historian of sound, Fitzhugh Green, whose 1929 study, *The Film Finds Its Tongue*, explains how operatic shorts often faced popular disinterest: "Audiences manifestly liked the vaudeville shorts better than they did the operatic ones, and Sam [Warner] and the Manhattan crew began making vaudeville acts and dance orchestras in preference to the heavier stuff."[21] As a result, a majority of the short-subject producers that made the move into sound for the 1928–1929 season opted to follow Vitaphone's Broadway strategy rather than its initial highbrow aspirations: MGM, for instance, had its *Metro Movietone Acts* (featuring "vaudeville stars or teams"), Universal its "vaudeville novelties," and Paramount an in-house series of *Paramount Talking Acts* ("produced with the cream of screen stars and of Broadway talent combined"), among many others.[22]

Nor was it only through canned variety and revue acts that the Broadway ethos left its mark on short subjects. Also significant were a number of dialogue-oriented shorts adapting theatrical sketches for the screen, and here, too, Warners was at the forefront. In June 1927, Warner Bros. had opened a new Vitaphone unit in Los Angeles, where supervisor Bryan Foy and writers Hugh Herbert and Murray Roth launched a series of *Vitaphone Playlets*—two-reel, all-talking adaptations of "refined" comic and dramatic sketches, of a style commonly featured on the legitimate stage and in big-time vaudeville. As Charles Wolfe has noted, Vitaphone's playlets were distinct from the more frontally staged vaudeville and presentation shorts, where the human figure was placed front and center before the camera; instead, the playlets turned the actors' performances "inward," harnessed to a self-contained fictional world.[23]

FIGURE 10. Advertisement for Vitaphone from *Photoplay*, October 1929.

But the playlets were also distinctive in proposing an alternative indicator of distinction within sound comedy from that offered by the presentation shorts. Whereas the latter often brought to the screen exponents of the lunatic vogue so prized by metropolitan tastemakers, the comic playlet testified to the lingering hold of an older, genteel model of plot-based "situation" comedy, for which sound was harnessed in the service not of cuckoo wordplay but of polite wit and character-based humor. Taking the baton, for instance, was Educational Pictures, which put a series of *Coronet Comedy* shorts starring Edward Everett Horton into production in late 1928. "Today, the all-talking Short Feature comedy has virtually brought a re-birth of humor on the screen," declared publicity for the third of these, *The Right Bed* (April 1929), describing the films as "farce playlets similar to the one-act plays seen for years in vaudeville."[24] More or less identical discursive strategies were also in play in promotion for the Christie Film Company, which entered talkie production around the same time with a prestigious distribution deal through Paramount. Long associated with the situation style of screen comedy, the Christie company soon began emulating the Vitaphone playlet as a way of consolidating its brand identity. As indicated by early publicity, the Christie talking shorts would include "short features adapted from stage plays" by well-known Broadway playwrights, produced under a policy of "cast[ing] them with stars from both stage and screen."[25] Scripted dialogue, in this conception, was to provide the royal road for a style of screen comedy that would forgo the vulgarities of physical knockabout, as studio head Al Christie himself explained:

> The field of comedy type of entertainment was limited before. After all, there were just so many different ways in which a man could be knocked down or lose his trousers and I think myself movie audiences were getting pretty fed up on this kind of striving for laughs. . . . [By contrast,] the new style of entertainment holds the audience interest far more. It has always good construction to get the interests of the audience by promising something and then working up to it. This can be done far better with the addition of good dialogue.[26]

Less predictably, these efforts also informed Christie's production of a series of six "negro stories" with all-black casts, adapting white southern writer Octavus Roy Cohen's "Darktown Birmingham" stories from the *Saturday Evening Post*. With a cast drawn from Harlem's Lafayette Players Stock Company, including Spencer Williams, the films corresponded to better-known shorts from this era in appropriating black performance to sound technology, such as Vitaphone's Expressionist-influenced *Yamekraw* (ca. June 1930) and RKO's Duke Ellington vehicle, *Black and Tan* (December 1929).[27] But they differed in recruiting black voices to the gentrification of sound comedy: rife with comic malapropisms and verbal misapplications, the films were sold to audiences in terms of the putatively literary pleasures of "original Negro talk" and praised for a "vocalized form" that may have correlated with the cuckoo taste for verbal nonsense.[28] (Certainly it was in cuckoo terms that one of the era's most acclaimed humorists, Richard Leacock, advocated the "higher ground" of what he called "Negro talk": "The humor of Negro talk moves on to higher ground when it turns not merely on sounds but on sense . . . and satirizes the Negro's fondness for long words, by which he confuses length of sound with depth of meaning.")[29]

The Darktown shorts may have departed from a Broadway model in the strict sense, yet they shared in the relationship linking sound's advent to the perception of slapstick's delegitimization: a straight line runs from Christie's disparagement of the "limited" appeal of people losing their trousers to the *Motion Picture News* critic who celebrated the Darktown series' success with audiences who "appreciate clean comedy without custard pies."[30] Already marginalized by the metropolitan trend of cuckoo humor, slapstick thus found itself downgraded once over by the ascendancy of the comic playlet and other sound-enabled comedic forms. The process again follows the model described by Bourdieu, who notes how any innovation within a given cultural field (e.g., sound) inevitably constructs a new polarization—what might be called "rehierarchization"—such that formerly dominant forms (e.g., slapstick) are demoted through a process of "social aging."[31] Indeed, in the case of slapstick, this social aging was immediately apparent as trade press articles began to ask questions like "Is old-fashioned slapstick to vanish?" virtually from the moment of sound's dissemination.[32] It was the opinion of Fitzhugh Green, for instance, that with sound, "slapstick passed out of date. It was too crude for 'pictures' that were acquiring tone and polish."[33] Such reports of slapstick's immanent doom were of course exaggerated, speaking rather to struggles over the definitions and hierarchies of comedy that sound made possible. Within the changing position takings that reshaped the field of comedy production following sound, slapstick had not so much begun its exit as become a kind of "low other" in relation to which new forms of screen comedy were being defined.[34] It is symptomatic, moreover, that assertions of slapstick's passing lasted only as long as sound's novelty permitted such new position takings. Once the initial waves caused by sound had been weathered, short-subject comedy at the major studios soon settled back into something like its accustomed slapstick form—even at Warners, where, as we have already quoted Peter Vischer's metaphor, "pies actually came into use again." In 1931, for example, Bryan Foy left the short-subject unit at Warner Bros. and was replaced by Sam Sax, under whose supervision the short-subject division returned to generically defined film series that, by the following year, included slapstick.[35] Nor was this in any way a surreptitious or sheepish reentry: the vaudeville and revue stars who quickly burned through their stage repertoire in sound shorts—discussed in the previous chapter—created a space into which veteran slapstick performers and filmmakers rushed to reclaim lost territory. Vitaphone's new series of *Big V Comedies,* for instance, boasted perhaps the most noteworthy coup: the return to the screen of former Keystone comedian Roscoe "Fatty" Arbuckle, signed to work at Vitaphone's Brooklyn studios following a twelve-year absence imposed by the MPPDA in the wake of an infamous 1921 scandal.[36]

But other studios similarly began to rehabilitate silent-era troupers around this time, albeit to less industry attention. Beginning in 1932, RKO substantially expanded its slate of live-action slapstick two-reelers, adding new series lines with star performers like Hal Roach stalwart Edgar Kennedy and, for the next season, erstwhile Sennett regulars Harry Gribbon and Tom Kennedy; Columbia Pictures did much the same, reorganizing its shorts division to become a home base for silent-era veterans like Charlie Murray, Andy Clyde, Harry Langdon, and many others.

Yet we will here briefly anticipate the theme of this book's final chapter by noting how slapstick's "return" was already, by this point, edged with nostalgia. The industry's Broadway strategy may have been short lived, but it rendered unmistakable the slapstick short's passage toward datedness: previous slapstick conventions were now cast as throwback comedy, its pleasures those of yesterday. Publicity for Arbuckle's Vitaphone shorts thus insisted that comedy had not changed since the "old days," as though the appeal of sound slapstick lay in its direct continuity with the gags and comic devices that had defined screen comedy two decades earlier.[37] The film industry's initial efforts to integrate the early talkie public around an assumed Broadway standard may not have eradicated slapstick, but it *did* align film industry practice with urbane hierarchies of evaluation in consigning slapstick to the temporal logic of the good old days.

"A STATE OF MORAL COLLAPSE": THE SHORT-SUBJECT INDUSTRY AND THE EXHIBITION LANDSCAPE OF THE 1930S

But the fact that it now had the irrevocable connotation of being "old-time" was only one consequence for the slapstick short of the industry's sound-era reshaping. Also relevant were contestations over exhibition practice that spoke to ambivalences around the industry's self-conception as a particular type of service. Here, we introduce what would become a new term in Hollywood's conception of its public as the industry now tested the cultural and political waters of the New Deal: the image of the filmgoer as a type of "citizen consumer." The concept of the citizen consumer derives from the work of historian Lizabeth Cohen, where it describes the rhetoric of civic-minded consumption and consumer rights that accompanied the economic reforms of Roosevelt's first term.[38] It was an ideal that was embedded in a number of the New Deal's keystone programs and acts, not least being the National Industrial Recovery Act (NIRA), passed into law in 1933, which enlisted consumer representatives as members of some of the code authorities and established a Consumer Advisory Board to give consumers a legitimate voice in the federal government's efforts to foster recovery. And it was an ideal that informed grassroots concerns about film industry business practices, too, spawning a series of contestations over Hollywood's role as a mass culture industry: as we will see, contentious exhibition practices like double billing, small-town audiences' complaints about movie morality, Hollywood's "Broadway strategy"—all of these were increasingly debated and discussed in a new language of consumer protection. "Have we one public or many publics?" was thus the telling complaint of one small-town exhibitor speaking for heartland audiences who resented the presumption that Broadway-style entertainment should constitute a national "mass" standard. (The essay was titled "Broadway and/or United States.")[39] From the perspective of consumer advocacy, culture industries like Hollywood were increasingly assessed and critiqued on a model of public service, for which the mass audience, to reverse one of Theodor Adorno's most famous barbs, was envisioned no longer merely as their object but ideally as their subject.[40] Controversies over film exhibition in this way became an opening through which a newly civic model of the film-going public began to be asserted, with significant upshot for the economic and entertainment function of short subjects on exhibitor programs.

To see how this came to pass, it will be necessary first to explore at some length the changing and contested place of "variety" as a film industry standard during the conversion period. By the late silent era, Richard Koszarski notes, "the experience of viewing a film [had become] far different from what it would be at any time before or since. Exhibitors considered themselves showmen, not film programmers." The feature film attraction constituted "only one part of . . . [an] evening's entertainment," which regularly included live stage presentations featuring professional dancers, comedians, operatic and popular singers, as well as short comedies, newsreels, and travelogues—all to provide the variety and heterogeneity that had long served as an American entertainment standard.[41] Silent-era moviegoing was thus, Koszarski suggests, essentially a theater experience rather than a film experience, inasmuch as live acts gave each show the irreducible singularity of a one-off performance.[42] Yet sound short subjects threatened to make the practice of live presentations obsolete. The very raison d'être of Warner Bros.'s pioneering Vitaphone shorts had been to provide smaller exhibitors with a cost-cutting substitute for presentation acts. Soon, industry insiders and commentators were voicing a death-knell chorus. As slapstick producer Jack White bluntly asserted in 1929, "Short dialogue comedies will kill the presentation racket. . . . [T]he thing is obvious, it speaks for itself."[43] In a similar spirit, Martin J. Quigley, then editor of the Exhibitors Herald-World, declared, "The short subject, with dialogue and music, . . . makes possible the return to an all-film policy which would not otherwise be possible. Pictures for picture houses is the best policy for the industry at large."[44]

One may of course wonder who Quigley had in mind as the "industry at large." For the truth is that these developments ultimately served the interests of producers far more than exhibitors: theater owners had in many cases embraced live presentations as a way of differentiating their shows from competing theaters, while the major studios had sometimes discouraged the practice because they diverted potential film rental revenue to live performers. Here, then, was a way in which the substitution of sound shorts for live acts could consolidate power in the producers' hands. But it was also a way of standardizing spectatorship and the viewing situation. The notion of spectatorial "distraction" so famously evoked in Siegfried Kracauer's description of silent-era movie palaces—when live performers, orchestral music, and lighting effects contributed to a moviegoing experience that dispersed spectators' attention throughout the space of the theater—soon became a thing of the past.[45] With the gradual disappearance of live acts, moviegoing became a more exclusively filmic experience: the three-dimensional space of the movie theater was now fully subordinate to the two-dimensional space of the screen; equally, the local and neighborhood orientation of much silent-era film exhibition, when theater owners had drawn upon regional networks of live entertainers, was increasingly constrained in the face of Hollywood's nationwide reach.[46]

These shifts in exhibition practice and programming did not, of course, come into immediate effect; elaborate stage revues continued to be booked as a mark of distinction for prestige houses into the early 1930s, and rural theaters featured live "hillbilly" musical acts through much of the decade.[47] That the writing was on the wall for live acts was nonetheless clear when Samuel "Roxy" Rothafel, the picture palace impresario most responsible for promoting blended programs of live and motion-picture entertainment, himself swore off the strategy in 1930, announcing to a group of Universal salesmen that "the day of merging the so-called presentation idea with the picture is past."[48] The paradox, however, was that sound short subjects were now looked upon to provide the very variety that they themselves had been largely responsible for taking away. Though programs of shorts had accompanied feature programs since the silent era, the changed circumstances of sound-era exhibition now caused exhibitors to pay closer attention to short-subject selection to ensure a balanced program in the absence of live acts. "The difference between the old silent shorts and the present-day talking shorts is almost like night and day," declared exhibitor Charles E. Lewis. "Previous to the sound era, shorts were better known as program fillers. Today, they are granted the more appropriate title of program builders."[49] The year 1931 seems, in fact, to have marked a dawning consciousness in this regard, as industry insiders and showmen began to advocate vocally for the value of varied bills of short subjects, using trade journals as a sounding board to share strategies for "plugging" their shorts. "Never in my long career as a theatre operator has the short subject been of such vital importance as it is right now," declared E. A. Schiller, vice president of Loew's. "The development of talking pictures . . . [has] raised the so-called 'shorts' to a program-importance they never had before."[50]

Yet far from pumping new life into the short-comedy field, these trends simply added competition. The silent-era dominance of slapstick shorts and newsreels was now roundly dislodged by increasingly varied classifications of short-subject genres, prompting Terry Ramsaye, in *Motion Picture Herald,* to enumerate with astonishment the "tremendous array of specialty products of appeal, travel, adventure, sport, historical and musical lore, compressed tabloid screen vaudeville, the delicious extravaganza of whimsical magic in the animated cartoon, personalities, nature studies, novelty in fashion and color, and serialized drama" in recent short subjects.[51] One symptomatic genre was the "musical revue," effectively a film substitute for a live prologue that, not coincidentally, thrived simultaneously with the prologue's decline. As exemplified by series like MGM's two-strip *Colortone Revues* ("Brilliant Tabloid Musical and Dancing Entertainment"), such films were musical shorts whose thin—often bizarre—narrative premises served simply as rationale for a series of spectacular musical performances, as in, for example, *The Devil's Cabaret* (December 1930), wherein Satan tries to make Hades a more appealing destination by putting on cabaret acts for its denizens.[52] Also typifying this rapid burgeoning of new genres was the sports short, a category that sprang almost from nowhere around 1930 to quickly become a mainstay of short-subject programs. The popular success of Pathé's six *Football with Knute Rockne* shorts in the 1930–1931 season prompted a land rush for other sports figures in instructional shorts, the most successful of which were golfer Bobby Jones's *How I Play Golf* and *How to Break Ninety* one-reelers for Warner Bros. (1931–1933). Such shorts were, moreover, microcosms of the strategies of diversified appeal that more broadly defined the short-subject industry during this period. The Bobby Jones series, for instance, typically involved Hollywood glitterati bumping into Jones on the course, thus combining sports instruction with "behind-the-scenes" peeks at stars at leisure (James Cagney, W. C. Fields, Edward G. Robinson, and Loretta Young all made appearances in the Jones series).[53] Others were even stranger hybrids, such as Vitaphone's 1930 bridge short *Milton C. Work, "The International Bridge Authority,"* which integrates sequences of bridge instruction into a domestic comedy about married couples quarreling over their card playing.[54] The quest for variety was, indeed, pursued with a delirium that bespoke a catch-as-catch-can approach to audience interests: outside of golf and bridge shorts, one now also found such series as MGM's *Fisherman's Paradise* ("depicting the mighty thrills of deep sea fishing," 1931–1932), Educational's short-lived series *As a Dog Thinks* ("human interest stories about dogs," 1933–1934), and RKO's *Dumb-Bell Letters* (one-reel compendia of odd letters received by American businesses, 1934–1935).[55] The ratio of live-action slapstick series to other short-subject lines released by the major studios immediately began to decline in the face of such diversification— from around half of all series released in 1930–1931 (not including newsreels) to around a quarter and dropping just five seasons later. Meanwhile, animation was enjoying an upward arc of popularity that made it by the mid-1930s the favored short-subject genre among general audiences—a position that it retained for the remainder of the studio era.[56]

Variety was no end in itself, however; it was also a tactic in the face of what quickly came to represent an unprecedented threat to the short subject's economic and industrial viability. I mean here to refer to the exhibition policy of double billing, which began to proliferate as one among a number of strategies introduced by theater owners reeling from the crisis of the Depression. (Others included prize games like Bank Night, which was a lottery system, and Screeno, a kind of bingo.)[57] The practice of exhibiting two features on a program was not unheard of during the silent era, when it was largely confined to theaters in the Northeast and Southwest; but it became widespread among Depression-era exhibitors, particularly independents, who sought to bolster dwindling box-office receipts by offering audiences more for their money. In 1931, some eighteen hundred theaters had instituted double bills, representing one-eighth of all theaters then operating in the country; five years later, however, *Film Daily* estimated that the proportion of theaters regularly featuring dual bills was now over half (eight thousand out of what was now around fifteen thousand movie houses nationwide).[58] The havoc wrought on the market for two-reelers was huge: with a double bill consisting of two full-length pictures and a newsreel, exhibitors were finding little room for two-reel films (which typically played from seventeen to twenty minutes). "The average two-reel comedy, they say, will be passed up for a one-reel cartoon, or some other subject which can be shown in less than 10 minutes."[59] As early as 1931, it was reported that lower rentals due to increases in double featuring were forcing short-subject producers to pare budgets to Poverty Row levels, with "the average spent on a two-reeler . . . around $25,000"—a drop of about ten thousand dollars compared to pre-Depression prices.[60] (In fact, the report likely underestimated the extent of budgetary cutbacks, at least to judge from those studios for which budget documentation remains extant. At RKO, for instance, production costs for most short-subject series were closer to twenty thousand dollars per picture during this period, while the company's popular Edgar Kennedy *Average Man* shorts were commonly brought in for a scant fifteen thousand dollars each. Columbia's short-subjects division similarly kept budgets to around fifteen thousand dollars per short throughout the 1930s.)[61] By 1935, the prognosis for short subjects had become even direr, prompting the *Herald* to anticipate that the "days of the two-reel comedy are fast drawing to a close."[62] Bold predictions of the "vital importance" of shorts quickly shriveled in the face of exhibition practices that rendered such importance moot.

The specific and varied strategies through which individual short-subject companies confronted this challenge will emerge in part 2 of this book. For the present, I want to simply indicate how industry figures who opposed double bills often made use of the rhetoric of consumer protection that was one of the hallmarks of New Deal–era economic reform—how, that is, the defense of variety programs comprising single features and shorts was mounted as a matter of consumer rights. There is no question that such appeals were in the main alibis for the economic interests of the film industry's power bloc: the vertically integrated majors viewed the "double feature evil" primarily as injurious to profit margins; high-end exhibitors and large theater chains similarly saw it as a matter of unfair competition and price gouging on the part of smaller houses. Much time and money was in consequence expended in the form of public surveys and advertising campaigns that sought to appropriate the rhetoric of consumer protection for the battle against double bills. A case in point was provided by one M. B. Horwitz, general manager of the Washington theater circuit in Cleveland and Cuyahoga Falls, Ohio, who was "ardently opposed to the double feature policy." In March 1932, he polled patrons at his flagship theater, The Heights ("one of the most representative first-run suburban picture houses in Cleveland"), for their opinion on double bills, claiming to do so in the name of the public's interest: "Rarely is the public—the final consumer—taken into consideration insofar as partic-ipation in these discussions is concerned. The distributor and exhibitor have adopted a 'know-it-all' attitude regarding double features."[63] The results confirmed Horwitz's own preference, with 76 percent of those polled voting for single features with an assortment of shorts. (Tellingly, the only exhibitors who demurred from Horwitz's subsequent Cleveland-wide petition against duals were "a few scattered theatre own-ers in remote sections of the city"—a likely reference to the smaller neighborhood houses, for whom double billing was a much-needed economic lifeline.)[64] A year later another interested party, Henry Ginsberg, general manager for the Hal Roach Studios, successfully lobbied the MPPDA to administer a nationwide questionnaire on the topic to organizations of "educators, women's club[s], . . . parents' associa-tions, editors, professionals, civic leaders and others"; when the results indicated an astounding 90 percent opposition to double features, Roach's publicity director, Lew Maren, aggressively telegrammed syndicated newspapers and trade publications to run the results.[65] Among those questionnaire replies that did see partial publication, notes were sounded that were already becoming commonplace in the arsenal of rhe-torical weapons against double bills: namely, that double features imposed mental strain, made family attendance impossible, and were out of step with moviegoer pref-erence. "It is tiring," "Double features are exhausting," "The mind is not refreshed, it is cluttered," "It leaves a confused impression," "It is too great a strain on the spectator," and "Double features are not relaxing" were typical replies, as was the insistence that "double features are objectionable to families who try to pick a 'family picture.'"[66] A Mrs. Thomas G. Winter of the Hays office summarized the questionnaire responses by insisting on theater owners' responsibility to their public:

> The double feature may have served a temporary purpose when there were myriads of people who were anxious to get as much as they could. But what is happening? The over-long program leaves them with headache, eyeache and a confused and tired memory to carry away, instead of a clear vision. . . . Many of our correspondents say they have given up going to the theatre rather than endure this over-long and incom-patible show, and particularly are keeping their children away.[67]

This kind of rhetoric only became more widespread in following years, mark-ing the emergence of a new construction of the moviegoing public that mirrored the larger reconceptualization of the role and rights of the consumer during the years of the Great Depression. Rumblings of consumer discontent in the face of producer interests had begun to be felt across all sectors of the economy as early as the mid-1920s, as expressed in best-selling books like Stuart Chase's *The Tragedy of Waste* (1925) and Chase and Frederick J. Schlink's *Your Money's Worth* (1927), but they greatly intensified as the Depression worsened, leading economists and social activists to call for new standards for consumer representation that would be enshrined in the economic philosophy of the NIRA. In the film industry, consumer discontent was hardly limited to debates about the double features, but registered across a number of areas—most impactfully, of course, in concerns about the movies' moral influence that resulted in the formation of the Production Code Administration in July 1934. That the double bill represented an early flashpoint in these processes is nonetheless of crucial importance for the concerns of this chapter, since it suggested a new way of framing the entertainment value of short subjects. Put simply: if duals were damaging to the public good (because they supposedly prevented family viewing, caused mental strain, etc.), then balanced programs of shorts and single features could be vindicated in the name of civic responsibility. It makes sense, then, that the resolutely anti-dual editor of *Motion Picture Herald,* Martin Quigley, regularly included in his publication the opinions of "public leaders" opposed to the practice—such as J. W. Hanson, principal of Roseville Union High School in California ("I favor one good feature supplemented by news reels, scientific reels, exploration, etc.") or Grace Morrison, president of the General Federation of Women's Clubs ("One *good* feature picture, a short comedy, a short travel or educational film and the newsreel gives to my way of thinking a properly balanced program"), or James E. West, chief executive of the Boy Scouts of America (who accused double bills of "preclud[ing] intelligent selection of one's entertainment" and eliminating "the motion picture as a feasible source of recreation for children").[68] We will see in the next chapter how one short-subject producer-distributor, Educational Pictures, launched a series of ad campaigns that similarly encouraged variety programming in the name of family values.

Yet the path to securing the continued viability of short subjects against double bills remained a fraught one. No amount of appeals to consumer protection seemed to abate the legal setbacks faced by the double bill's opponents, and by decade's end, the institutional framework on which the short subject's economic viability depended was in tatters. Despite sustained lobbying against the practice on the part of the vertically integrated studios as well as the Motion Picture Theatre Operators of America, the National Recovery Administration (NRA) officially legalized double bills when the Code Authority for the motion picture industry addressed the issue in August 1934—a ruling that was seen as a victory for the "little fellow" (i.e., independent theater owners).[69] The NRA also weakened the major studios' ability to strong-arm bookings of their shorts. Whereas previously exhibitors had often been forced to accept many more major-studio short subjects than they could reasonably play—a form of block-booking known as "full-line forcing"—the Motion Picture Code now conceded a more equitable arrangement: distributors could only force shorts in proportion to the number of rented features.[70] Nor did the vertically integrated majors gain headway when they began to introduce clauses in exhibitor contracts to prevent their films from being screened as part of double features. In late May 1934, Harry and Louis Perelman, owners of two independent theaters, filed suit in the US District Court in Philadelphia against all companies using anti-dual clauses, charging "distributor discrimination in the enforcement of anti-double feature clauses."[71] When the judge eventually sided with the plaintiff that December, the major studios responded by appealing to the US Circuit Court of Appeals, which nonetheless upheld the Philadelphia court's ruling in January 1936. The studios continued to appeal—MGM most stubbornly—but by 1938 the writing was on the wall: as exhibitor Frank H. "Rick" Ricketson admitted that year in *The Management of Motion Picture Theatres*, "The double bill, as an evil, . . . cannot be eliminated by distributors' contracts, exhibitors' arbitration boards, new codes, or any other route that has yet been explored."[72] The marketplace upon which short subjects competed was thus, by the mid-1930s, one that had seen the concept of variety programming change from industry standard to exhibitor choice. The legacy of changing exhibitor practices in the 1930s was to have left the short-subject industry in what one critic diagnosed as a "virtual state of moral collapse."[73]

"SOMETHING TO THINK ABOUT": THE EDUCATIONAL USES OF SHORT SUBJECTS

What emerges from the foregoing section is a twofold change in the rhetoric used to advocate variety programming during the 1930s. First, there was a shift in the meaning of variety from the "virtual Broadway" paradigm of the early sound era to a new conception that linked variety to ideals of consumer protection. Second, accompanying this, there was an emergent change in the industry's construction of the public for its films, from an object of the dissemination of metropolitan culture to a civic public of citizen consumers. (The two changes were, in fact, strictly correlated.) This emergent rhetoric admittedly had some troubling implications for slapstick producers, since rough-and-tumble physical comedy was not the kind of entertainment upon which claims to civic responsibility could readily be founded. (When "public leaders" like principal J. W. Hanson advocated the value of shorts, they did not have the Three Stooges in mind!) The short-subject industry's ongoing bid for validation thus had the effect of reframing the marginalization of slapstick—no longer, as in the immediate period following sound, judged inferior in relation to a Broadway style of comic playlets and metropolitan wit, but rather out of fit with the privileging of what the *New York Times* called "educational and instructive" shorts.[74]

All of these themes played out loudly in what became something of a last-ditch effort on the part of a major studio to create a renaissance in the short-subject field.

It was no coincidence that the studio in question was the very one that had struggled most stubbornly against duals: Loew's-MGM. No coincidence, too, that the company launched an ambitious revamping of its shorts selection almost simultaneously with the failure of its appeal against the Perelman ruling in January 1936. By this point, Hal Roach was winding down his distribution arrangement with MGM, and the studio took the opportunity to radically rebrand its shorts. Under Jack Chertok, head of the shorts division since 1935, MGM now burnished its short-subject offerings by introducing a growing number of what it described as "informative" product lines: the long-running *Crime Does Not Pay* reenactments of true crimes (1935–1947), *Historical Mysteries* on topics such as the fate of the *Mary Celeste* and the escape of John Wilkes Booth (1937–1938), Carey Wilson's *What Do You Think?* pictures investigating psychic phenomena (1937–1941), as well as a *Soldiers of Peace* miniseries celebrating medical improvements that have "made this world a safer, healthier, happier place in which to live" (1938).[75] Even comic shorts were conscripted to this instructional rhetoric, albeit in motley, in the form of the popular *Pete Smith Specialties* (1936–1955), in which Smith lent wry commentary to documentary topics of wide-ranging general interest, and the Robert Benchley "How to" series (1935–1940, 1943–1944) on the frustrations of daily life. Slapstick, meanwhile, was reduced to a single product line, the *Our Gang* series, now one-reelers, which MGM took over from Roach upon the latter's departure. To read MGM's own press releases, however, one would have imagined that slapstick had been entirely excised—so frequently was the form evoked as the evolutionary lower rung against which Chertok's more educational aspirations rebounded. As the studio's own bimonthly short-subject guide, *MGM Shortstory*, confidently declared in 1939: "Pointing to the strides made in the past year or two, MGM believes that shorts, finally rid of the stigma of slapstick, are in a stronger position today than at any time since the advent of double features"—despite the fact that the studio was still offering twelve *Our Gang* one-reelers every year.[76] "Let's keep our shorts educational," enjoined critic Allen Saunders, in an editorial consigning slapstick to an anticipated future as a museum relic. "Let's leave the custard pie, the philandering husband situation and the pratt fall to the movie museum" (fig. 11).[77]

Once again, a rhetoric of betterment informed these efforts, as with Vitaphone's first sound shorts a decade earlier; only now that rhetoric was shaped not by discourses of cultural taste and distinction but by social democratic notions of what film scholars Haidee Wasson and Charles R. Acland have termed "useful cinema."[78] The analogy that Chertok liked to draw for MGM's shorts was thus not with Broadway sophistication but with nonfiction literature and magazines. "Shorts are to the motion picture audience what non-fiction material is to the readers of magazines. . . . They are informative, and some of them have a moral. Although they are told as entertainingly as possible, they leave the audience with something to think about."[79] "I don't know about the public of 1928, or of 1918,"

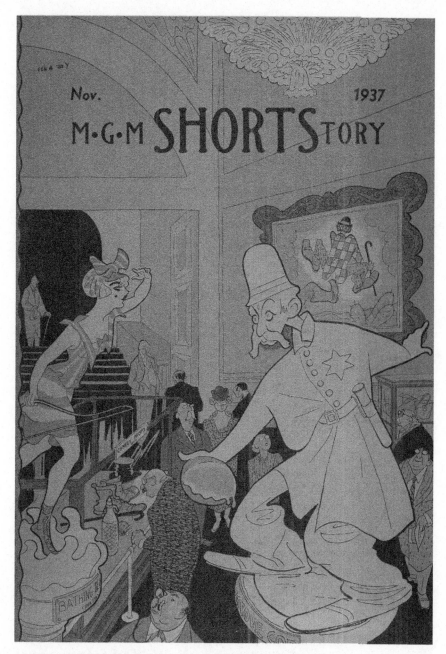

FIGURE 11. The cover of the November 1937 issue of *MGM Shortstory* envisions slapstick as a museum exhibit. Courtesy Billy Rose Theatre Collection, New York Public Library.

Chertok claimed around the same time, "but the public of 1938 likes to extract information from short subjects."[80]

Chertok's commitment to "informative" subjects on historical and moral themes corresponded to what was an emerging golden era in the classroom use of film, in which commercial short subjects played an unexpectedly prominent role. A growing body of recent scholarship on educational film has shown that the sites of film viewing significantly expanded through school classrooms in the 1930s as progressive educators seized upon film's promise as a means of implanting the citizenship skills needed to sustain democracy.[81] Of course, such initiatives were fueled foremost by specialist companies and public agencies; yet the commercial industry found reason enough to accommodate itself to these endeavors, too, particularly following the public relations crisis precipitated by the censorship debates of the early 1930s.[82] The MPPDA had made initial gestures toward educational goals in 1931, when it sponsored the formation of the Committee on Social Values in Motion Pictures, which produced the *Secrets of Success* series of twenty short subjects "about interesting people and how they behave."[83] These efforts were subsequently strengthened following the movie morality protests of 1933–1934, at which point the MPPDA's attention fell firmly on commercial shorts as the ideal format for burnishing the industry's pedagogical commitments: unlike features, shorts could be screened in schools with minimal impact on theatrical box office and were, furthermore, tailor-made to the constraints of classroom schedules. Accordingly, in September 1936, the MPPDA established an Advisory Committee on the Use of Motion Pictures in Education, whose members began a well-publicized process of reviewing noncurrent shorts with a view to building an inventory for school use. Outside of the Advisory Committee's efforts, the MPPDA also worked with the Commission on Human Relations film project of the Progressive Education Association, which, under the direction of Alice Keliher, similarly sought to develop a catalog of short subjects and truncated features, to be used for what Keliher described as "the study of our problems of human relationships."[84]

It is against the context of these initiatives that developments in MGM's short-subject lineup need to be situated. Certainly, no series of shorts was so frequently singled out for pedagogical value as MGM's. The studio's *Crime Does Not Pay* line received particular praise, in fact, for two episodes on the dangers of driving under the influence: *Hit and Run Driver* (December 1935), which Keliher's Commission on Human Relations included in its catalog to illustrate the "human problems involved in drunk driving," and *Drunk Driving* (October 1939), which prompted an approving letter to Chertok from a New Orleans pastor of the First Baptist Church.[85] Keliher also reserved special approval for the *Soldiers of Peace* series, using one of its entries as self-explanatory justification of her commission's faith in the educational possibilities of commercially released films: "There is no reason why films being shown currently in the theatres . . . should not be used for discussion by schools, clubs, and study groups of all kinds. Why should not the

MGM short, 'That Mothers Might Live,' [April 1938, on Dr. Ignaz Semmelweis's fight against puerperal fever] be the subject of discussion for high school and college science and hygiene classes?"[86] Nor was MGM at all reluctant to boost these educational uses. "We don't make any of our pictures directly for school children," Chertok once admitted, adding, however, that "in every short subject we try to inject some educational value, and all may be seen by children of school age with advantage."[87] Indeed, the *MGM Shortstory* at times read more like an educational brochure than an entertainment catalog, featuring articles and letters penned by notable progressive educators, including Keliher herself, Cleveland Public Library director Marilla Waite Freeman, and the motion picture chairwoman of the Texas Federation of Women's Clubs, Marietta Brooks. (Even J. Edgar Hoover contributed a piece—titled "Combating Crime through Movies!"—in praise of the *Crime Does Not Pay* series.)[88] Together, these and other writers helped craft a new narrative of the short subject's evolution, one whose path was expressed not in terms of the language of cultural distinction and hierarchy—as with publicity for the early Vitaphone shorts—but rather as a coming-of-age tale that emphasized civic ideals of instruction and intelligence. MGM, one *Shortstory* contributor noted, has "speculatively [begun] to put a few intelligence-vitamins into the diet of that long under-nourished and undeveloped specimen—the short subject.... Lavished with attention and treated with intelligence, the short, once an orphan, [has] stepped out to become the industry's fair-haired child."[89] Added another: "The short subject has metamorphosized [*sic*] from a comic section to a short story. In short, the short subject is growing UP!"[90]

Was MGM sincerely seeking an educational address for its shorts? The school market could hardly have represented more than a negligible source of profits for a company of MGM's stature. (Films released through the Advisory Committee—renamed Teaching Film Custodians in December 1938—were sold at a rental rate of fifteen dollars per year or thirty dollars for three years.)[91] Yet symbolic importance here outweighed any genuine pedagogical commitment. What mattered was the public goodwill that the appearance of such civic-minded endeavors could vouchsafe, as well as the implied justification of the shorts' value as program builders. The movie morality debates may well have been weathered, but the film industry remained buffeted during this period by waves of bad publicity requiring ongoing public relations efforts: the Justice Department's continued investigations of the studios for antitrust violation, exhibitor complaints about stars who were deemed "poison at the box office," trade press reports describing "public nausea" and "nationwide ... pessimism about the movies"—as historian Catherine Jurca notes, this was an array of problems unmatched even by the decade's earlier controversies.[92] Previous contretemps over double-billing and immoral films may have been flashpoints for an emerging rhetoric charging the industry with irresponsibility, but by the mid- to late 1930s, these rumblings reached a crescendo that forced industry-wide emphasis on ideals of public service and civic edification, of which

MGM's rebranded shorts lines were but one symptom. Elsewhere, the industry's newfound civic credentials were loudly proclaimed during an unprecedented 1938 public relations campaign, titled "Motion Pictures' Greatest Year," in which the industry aggressively sought to countenance public disenchantment by promoting the idea that it was now the values of the "average movie-goer" or "Joe Doakes and his girl" that were the industry's mandate—as though the public was itself a collaborator in the industry's success.[93] Within the industry's feature film output, meanwhile, a democratic commitment was also palpable in what Jurca has provocatively characterized as the "death of glamour"—a decisive shift in long-standing Hollywood publicity practices that had formerly promoted stars as "idols of consumption" toward a "just folks" approach that presented stars as mirrors of the general public and featured them in family-type films.[94] In all these ways, the urbane self-image which the industry had initially projected in the early sound era was ceding to its "populist self-fashioning" as a force for civic cohesion.[95] The story of Hollywood's changing conception of the mass public over the course of the 1930s thus pivoted around a rhetorical switch in Hollywood's position within long-standing divisions between cultural elitism and cultural populism: a mode of mass cultural organization formerly organized on a logic of *difference* (on a hierarchizing rhetoric of metropolitan sophistication) was being replaced by a more civic-populist logic of *equivalence* (predicated on an assumed correspondence between industry product and the entertainment needs of the "ordinary" citizen).[96]

Still, it was arguably the short subject that provided the most sustained model for these new types of address, the ideal low-cost test balloon for projecting the industry's claims to civic-mindedness, and not only at MGM. Writing optimistically about the "Two-Reeler's Comeback" in 1941, *New York Times* critic Bosley Crowther offered a survey of "adult content" in shorts as a broad-based trend that promised better fortunes for the short-subject industry as it entered the new decade.[97] He pointed to a more informational style of travelogue, such as travel writer Lowell Thomas's *Going Places* for Universal, which no longer offered merely "picture postcard" surveys (a "sort of review of monuments and mausoleums which invariably did a sunset fade with 'and now the time has come to say farewell to old Name-Your-Country'") but rather expressed "political and economic point[s] of view." He pointed to the launching in 1935 of *Time* magazine's *March of Time* newsreels, distributed by RKO, as an "experiment in cinematic journalism"—a "sort of editorial newsreel with perspective" that had become "one of the most popular and influential fixtures on the screen." And he lent his voice to the chorus of praise for MGM's *Crime Does Not Pay* series, which had already spurred imitators at other studios (such as Universal's one-off three-reeler *You Can't Get Away with It!* [November 1936]).[98] "As a direct result of these and other more recent efforts," Crowther concluded, "shorts today are indisputably superior in quality to any that have gone before. More money and brains are being devoted to them, and they are of infinitely greater variety. . . . It has been a long road for shorts, with a stretch through a dark valley. But now they're on rising ground."[99]

"THE ZANY CREATURES THAT PEOPLE THIS EARTH": NEW DEAL-ERA POPULISM AND THE COMEDY SHORTS OF ROBERT BENCHLEY

Yet the renewed emphasis on instruction is only one way civic ideals colonized the short subject during these years. I therefore want to return to the question of comedy to draw some observations about the links binding changing forms of Hollywood's self-presentation to associated forms of comedic representation. It is no doubt a fluke of history that a single humorist would play an important part in short-subject comedies across each phase of the developments so far discussed in this chapter; still, it is a fluke that sheds crucial light for tracing these links. The humorist in question was *New Yorker* writer Robert Benchley, who was first brought in front of the motion picture camera in 1928 to film *The Treasurer's Report* (ca. May 1928) for the Fox Film Corporation, but subsequently went on to win great acclaim for his "How to" series for MGM, beginning in 1935 and lasting through 1944.[100] Over the course of the 1930s, I want to show, Benchley's comic persona changed in ways that corresponded to Hollywood's New Deal–era civic refashioning. It was a change that first registered when the former *Life* drama critic took up residence at Chertok's rebranded short-subject unit, first in a one-off, *How to Sleep* (September 1935), released through the *MGM Miniatures* line, and subsequently in his own series; and it was a change that was often remarked as Benchley continued to broaden his media presence across the decade. By the end of the 1930s, in addition to his MGM shorts, his regular magazines and newspaper columns, and his frequent collections of essays, Benchley also hosted his own nationwide radio show on CBS, *Melody and Madness* (making him the sixth most popular radio personality on the air, according to a *Radio Daily* poll), and lent his likeness to newspaper and magazine advertising campaigns promoting everyday goods like General Mills food products and Serta mattresses.[101] The change in Benchley's persona, furthermore, was always described in the same way, as a passage from Benchley's early reputation as a "smart" literary wit in the 1920s to a new, mass media identity as a Depression-era "average man." As Chertok himself put it, once Benchley had been "just a comic fellow appealing to sophisticated audiences"; now he was "becoming the average man to the public, which extends all over the country."[102] In a career that embodied the trajectory describing the changing dynamics of American culture, Benchley's initial reputation as a "smart humorist" thus began to shade into a new identity as a "down-to-earth humorist," his early fame as among the era's most celebrated New York wits now leavened with a more populist appeal, in line with the film industry's own "just folks" civic appeasements.[103]

Benchley's "becoming average" partook in a broader preoccupation with "averageness" and the values of the "ordinary" citizen that provides one key to understanding the era's changing cultural politics. The construction of identities in support of the New Deal had required a political language capable of bringing a heterogeneous social reality toward an imaginary equality; it found this in the concept of the "Common Man" and its cognates ("Joe Doakes," "John Q. Public," etc.), all of which became synecdoches for a national culture held in common. Historian Warren Susman pinpointed the 1930s as a time when many in the United States made the effort to characterize and adapt to a shared "American Way of Life," as scholars and intellectuals from Constance Rourke to Van Wyck Brooks sought in their writings to explain the significant cultural and historical values, experiences, and attitudes that the nation's citizenry was said to hold in common.[104] It was during this period, too, that developments in the social sciences—notably, the launching of George Gallup's American Institute of Public Opinion in 1935—gave credence to the idea of a "typical" or "average" American as a cultural arbiter lending imaginary stability to the period's social upheavals.[105] The "average American" was in this sense at once a statistical guarantee and a normative ideal for a nation committed to the recovery of a basic unity. Yet what was further at stake in the particular case of Benchley—and what makes his evolving persona more than just a symptom—was the role humor played in modeling this commitment. The passage from "sophistication" to "averageness" as terms of Benchley's appeal corresponded not only to broader patterns of cultural discourse but also to competing modes of rhetoric in Benchley's humor in the years immediately before and after twentieth-century capitalism's greatest crisis—on the one hand an absurdist rhetoric that enacted forms of semantic and symbolic distinction (under the rubric of sophistication), on the other a populist rhetoric that withheld such differentiations (under the rubric of averageness). What the trajectory of Benchley's development can ultimately provide, then, is a case study of how Hollywood's participation in the changing coordinates of Depression-era mass culture might further be unpacked through an analysis of comedic form, that is, of technical features within the field of comedic expression whose implications nonetheless extended beyond that field.

Few humorists had fared as well as Benchley on the waves of Hollywood's initial appropriation of metropolitan culture to sound cinema. He was not the only *New Yorker* wit recruited to shorts during the conversion period—Donald Ogden Stewart, for instance, starred in a couple for Paramount in 1929, *Traffic Regulations* (ca. February 1929) and *Humorous Flights* (April 1929), and Alexander Woollcott would later show up in the novelty short *Mr. W's Little Game* (June 1934)—but Benchley was the first, and it was the critical acclaim of his early film appearances that made possible his colleagues' later, more poorly received shorts. Benchley's debut, *The Treasurer's Report,* is in fact of no small historical interest in that it preserves an extremely popular live sketch he had first developed in 1922, when he and members of the Algonquin circle staged an amateur revue entitled *No Sirree!* A parody of a nervous speaker, the one-man sketch featured Benchley as the treasurer of a Kiwanis-type civic organization who is asked to give the group's financial summary. Despite intending it as a one-off, Benchley was subsequently invited to perform "The Treasurer's Report" in Sam Harris and Irving Berlin's *Music Box Revue* of 1922–1923. He next went on to tour the routine at vaudeville houses and, finally, submitted to Fox executive Thomas Chalmers's requests to perform it before the Movietone cameras at the studio's Astoria sound stages. As Benchley later recalled:

> I guess that no one ever got so sick of a thing as I, and all my friends, have grown of this Treasurer's Report. I did it every night and two matinees a week in the Third Music Box Revue. Following that, I did it for ten weeks in vaudeville around the country. I did it at banquets and teas, at friends' houses and in my own house, and finally . . . made a talking movie of it. In fact, I have inflicted it upon the public in every conceivable way except over the radio and dropping it from airplanes.[106]

The filmed version of "The Treasurer's Report" received extraordinary critical acclaim for a short, prompting some critics to write as though the film single-handedly justified the coming of sound. "[Mr. Benchley] is, by all odds, the best excuse for the talkies that has been yet invented" was the opinion of one critic.[107] "Robert Benchley Shows That All the Talking Pictures Need Is TALENT," yelled a headline in *Screenland,* in preface to a full-page transcription of the film's monologue.[108] *Variety* meanwhile observed that the film "has scored more laughs than anything ever turned out in talking shorts."[109] That a one-reel subject could thus be represented as the talkies' salvation should not, however, be surprising: Benchley's reputation as the vanguard of urbane literary humor was perfectly suited to the film industry's strategies of metropolitan distinction during the conversion period. Fox subsequently capitalized on Benchley's debut by signing him for five more shorts at five thousand dollars apiece, this time to be filmed in California, where he completed another satire of public speaking—*The Sex Life of the Polyp* (July 1928), in which he awkwardly discusses the titular creature's mating habits at a women's luncheon—and then a number of more conventionally plotted short comedies.

Benchley's initial film appearances did not, however, extend beyond the period of the talking picture's novelty. Benchley's relation to the industry remained at this point that of an East Coast outsider who, in 1931, apparently stirred executives' ire by describing Hollywood as "a flat, unlovely plain, inhabited by a group of highly ordinary people" and "the dullest and most conventional community of its size in the country."[110] During the first half of the 1930s, in fact, Benchley largely withdrew from film work, his Hollywood endeavors now limited to a handful of cameo roles, occasional script doctoring, and a one-off short subject for Universal, *Your Technocracy and Mine* (April 1933), in which he again performed his awkward lecturer shtick. Regardless, his debut appearances remain as luminous examples of early sound Hollywood's endeavor to harness what Gilbert Seldes dubbed the "cuckoo school" to its metropolitan rebranding, case studies for a form of humor predicated, as we have seen, on the art of "suspend[ing] the fancy between . . . incompatible [meanings]."[111] Take, for example, an excerpt from *The Treasurer's Report* in which, after some awkward initial pleasantries, Benchley's treasurer begins to race through his financial statement:

> During the year 1926—and by that is meant 1927—the, er, Choral Society received the following in donations: BLG—five hundred dollars; GKM—five hundred dollars; Lottie and Nellie W.—two hundred dollars; "In memory of a happy summer at Rye Beach"—er, ten dollars; proceeds of a sale of coats and hats which were left in the boathouse—fourteen dollars and, er, fourteen dollars. And then the Junior League gave a performance of "Pinafore" for the benefit of the fund which, unfortunately, resulted in a deficit of three hundred dollars. Then we took in from dues and laboratory fees $2,345 and fifty-five, no, seventy-five cents—er, making a total of receipts amounting to $3,645.75. This is, of course, all reckoned as of June.
>
> Now in the matter of expenditures, the, er [*Benchley distracted by waiter clearing up in front of him*]—in the matter of expenditures, the club has not [*clears throat*] been so fortunate. There was the unsettled condition of business and the late spring to contend with, er, resulting in the following rather discouraging figures, I am afraid. Er, expenditures $20,574.85. Then there was a loss, owing to several things of $3,326.80, carfare $4,452. And then Mrs. Crandall's expense account, when she went to Baltimore to see the work they are doing there, came to $119.50, but I am sure that you will all agree with me that it was worth that to find out, er, what they are doing in Baltimore. . . .
>
> Now, these figures bring us down only to October. In October my sister was married, and the whole house was torn up, and in the general confusion, er, we lost track of the figures for May and August.

What we can begin to detect in *The Treasurer's Report* is what one literary scholar of the time celebrated as Benchley's quality of "slightly made inconsecutiveness . . . the humor of the incongruous and the inconsecutive carried to its nth power," here manifest as a strategy of *enumeration*.[112] The humor here resides not in the simple fact that the math is wrong, but rather in the way the lecturer's enumeration finds itself stretched across a heterogeneous series of categories that don't quite "agree": a *reason* for a donation is listed as though it were the *agent* of the donation ("In memory of a happy summer at Rye Beach" included as though of the same ontic category as the named donors "BLG" and "GKM"), a source of loss is introduced as a kind of negative or inverted profit (the Junior League's benefit performance), the chaos of a wedding party the cause of disappearing numbers. The system of language, in its headlong flow, here becomes the element in which Benchley's narrator tries to hold these dissymmetries together, his rush of words trying to ward off the presentation's dissolution into the free play of nonsense. Here, one might say, nonsense is on the horizon as something to be

FIGURE 12. Robert Benchley's "informational" chart in *Your Technocracy and Mine* (April 1933). The annotations read, left to right, "Saturday night," "Monday morning," "Site of new country club," "Way down below site of new club," "Thermodynamic arrivation," "Himmelwaldsee," and "114."

guarded against, an ever-present risk that discomfits the lecturer's presentation; in other shorts from this early period, however, nonsense more notably overtakes matters, a case in point being Benchley's one short for Universal, *Your Technocracy and Mine*, in which he attempts to explain the Depression-era social-engineering buzzword of the film's title. (The opening intertitle situates the film in the context of debate over this widely satirized concept: "A few more words to add to the general confusion by an expert who isn't quite sure about the whole thing himself.") Here, the enumeration is woven around a line graph that evades the lecturer's ability to pin it to any clear signification (fig. 12). To quote from the monologue:

> Now this ought to make it a little clearer what I'm driving at, and ought to bring home to you the importance of the situation to you. This chart represents the output of energy of one man over a period of one year. Here [*points to first peak*] is the energy available on Saturday night, and here [*points to first trough*] the energy available Monday morning. These are all, of course, figured in energy-determinants—as here [*third peak*] you will see thermo-dynamic arrivation. Thermo-dynamic arrivation means, er—well, you know what thermo-dynamic arrivation means. Er, now here [*second peak*] is where— is the site of the new country club where the lockers and showers are going to be.

And down here [*second trough*] is a place way down below the site of the new country club. All of this here is made land—the river once came all over this. Now from here [*third peak*] we went down the mountain [*moves pointer down to third trough*] to a beautiful little town in the valley named Himmelwaldsee, where we rested after our long walk and partook of the cream and cakes with which that section of the valley abounds. There's no place to go from here [*fourth peak*] so we—no sense in going way up here—we might just get stuck here for the rest of the night, so we won't do anything more with that. Now here [*points top left*] is—represents—the horse-power available in the State of Ohio alone; this is just a front view, you see. Er, over here [*points top right*] seems to be a loose number that doesn't have any connection with anything. I think that's just 114 and we'll just call that 114 and let it go at that.[113]

The line graph is first read by Benchley as a symbolic representation plotting a straightforward variable ("the output of energy of one man") against scientific-sounding nonsense ("energy determinants"/"thermo-dynamic arrivation"). Yet the lecture comically switches to an iconic register of interpretation, whereby graph space is literalized, first as geographic cross-section ("down the mountain"), next as perspectival representation (a "front view" of horse-power in Ohio). The absurdist exercise here consists in forcing incommensurable systems of signification (symbolic/iconic) into an impossible dialogue (graph/cross-section/view): like the "loose" number "114," all signifiers in such a situation become floating signifiers from whom the very possibility of meaning is deferred.

The "incompetent lecturer" model in these and other of Benchley's early shorts has precursors in earlier traditions of humor. Norris W. Yates's critical analysis of Benchley charts a connection linking his early short subjects to the late nineteenth-century practice of the comic lecturer, as exemplified in a number of so-called literary comedians who created naïve or foolish personas for their writings and performances—for instance, David Ross Locke in his alter ego as Petroleum Vesuvius Nasby.[114] But Benchley's specific device of the comic enumeration, particularly in *Your Technocracy and Mine,* might also remind us of Michel Foucault's famous opening to *The Order of Things* (1966), where the author quotes a passage from Argentine writer Jorge Luis Borges that describes a "certain Chinese encyclopedia" in which animals are classified into the following groups: "(a) belonging to the Emperor, (b) embalmed, (c) tame, (d) suckling pigs, (e) sirens, (f) fabulous, (g) stray dogs, (h) included in the present classification, (i) frenzied, (j) innumerable, (k) drawn with a very fine camelhair brush, (l) *et cetera,* (m) having just broken the water pitcher, (n) that from a long way off look like flies."[115] "The monstrous quality that runs through Borges's enumeration," Foucault adds, "consists . . . in the fact that the common ground on which such meetings are possible has itself been destroyed. . . . Where else could they be juxtaposed except in the non-place of language?"[116] Mutatis mutandis, might one not claim that the bewildered quality that runs through Benchley's incompetent enumerations resides in the fact that any meaningful key to make sense of the data has gone missing? In such a circumstance, it is only within the "non-place" of a system of language, whether in the abstract space of a visual graph that lacks a key or in the rush of words through which Benchley's lecturer struggles to keep one step ahead of the collapse of his own presentation, that these data can be juxtaposed and combined. The very act of enumeration achieves a power of "mad inconsecutiveness" all its own, unfurling a series of errant signifiers that fail equally to designate a broader concept (e.g., technocracy) or to clarify an actual state of affairs (e.g., a club's finances) and that, as such, obey the absurdist logic of untethered sense.

But the "cuckoo" Benchley who first appeared in shorts in the late 1920s was not quite the "average" Benchley who would rule the MGM short-subject roost through the latter 1930s. Here, it is vital to reemphasize that the years of Benchley's expanding media presence in the 1930s were also years in which the cultural politics of metropolitan exclusivity were chastened by a new commitment to a kind of civic populism—that is, to the ideal of a "people's culture" whose potent framework was provided, we have seen, by a new emphasis on the rights and needs of the "ordinary" citizen. And ditto Benchley, whose humor now exemplified a displacement of "good" taste as a term of his earlier appeal toward a configuration of "mass" taste that invoked an imagined American commonality.[117] Three significant shifts, accordingly, distinguish the newly minted populism of his lecturer persona. First, in pace with developments in his writing, there was now a move away from pseudo-scientifically oriented topics (*Sex Life of the Polyp, Your Technocracy and Mine*) toward daily routine and leisure (e.g., *How to Start the Day* [September 1937], *How to Raise a Baby* [July 1938], *How to Watch Football* [October 1938], etc.). Not that this was a new direction for Benchley, who had long specialized in exasperated accounts of quotidian frustrations; what was distinctive was the degree to which this emphasis now saturated his film work, while his higher flights of absurdism were consigned to the more limited public for his writings. Second, whereas his first Fox shorts had him lecturing to a diegetic audience, the MGM shorts have Benchley giving his lectures direct-to-camera—that is, directly to the filmgoer—a formal switch that abstracts from the exclusive soirees and after-dinner speeches of his earlier appearances. Third and finally, the films now include comic visualizations of the situations that the Benchley lecturer describes, with Benchley himself appearing in them as their put-upon protagonist (commonly under the character name Joe Doakes, described by one critic as "a typical, good-natured, credulous, often-blundering American").[118] In place of the earlier stratification of social roles dividing Benchley from his audience (both the diegetic audience and the implied viewer), there is now a more complex structuring that collapses stratification into an assumed identity: Benchley is presented not only as a lecturer but also as somebody who experiences the same frustrations that he attributes to his audience. The populism of Benchley's address thus involves the discursive positing of some annoyance that serves as a

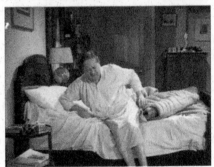

FIGURES 13–15. Benchley's MGM films typically feature a direct-to-camera presentation, in which Benchley himself appears both as lecturer on life's petty annoyances (13) and, in the accompanying comic visualizations, as fellow sufferer (14, 15). Frame enlargements from *How to Sleep* (September 1935).

term linking Benchley (whose frustrations are depicted in the comic visualizations accompanying his lecture) to the implied audience (whose same frustrations are assumed by the lecturer). Not only, then, are his screen lectures *about* "the customs and habits of Mr. and Mrs. Everyday America," but he himself *becomes* an everyman as his audience's onscreen surrogate (figs. 13–15).[119] What becomes primarily important to the rhetorical operations of his comedy is, then, no longer an absurdist enumeration of incompatibles but a recognition of shared identities.

"The popularity of his shorts," one magazine profile explained, "is doubtless due to his formula of placing himself in situations which everyone in the audience has already experienced."[120]

But it is not only the form of his address that has changed here, but also the very modality of humor within the shorts. Benchley may have unwittingly put his finger on the change when he described his new "average man" persona in a 1939 interview, shortly after the launching of his nationally syndicated CBS radio variety program. "'It is not a question of being a mirror of the times,' Mr. Benchley explained, 'but of holding a mirror to nature and permitting man to play the clown he so frequently is. . . . I had to annihilate the specter of smart humorist that trailed me and, in a nutshell, *be just myself or any other of the zany creatures that people this goodly frame, the earth.*'"[121] The key term here is *zany*, as a category of comic behavior that literary theorist Sianne Ngai, in *Our Aesthetic Categories*, has linked to the permanent mutability and role playing that defines labor in modern capitalism—and that Benchley's comment here posits as the very cornerstone of a shared populist identity in Depression-era America.[122] "All Aboard for Dementia Praecox" was thus the rallying cry for a comedic manifesto published by Benchley in 1934, for which he adopted the pose of a broken-down everyman plagued by "defective judgment," "retarded perception," "restrictions in the field of attention," "lack of motor skill," and "stupor."[123] What seems significant here is the way Benchley's zaniness thereby anticipates a performance mode that, according to Ngai, would flourish primarily in later decades—that strained style of incessant doing that links, say, Lucille Ball in *I Love Lucy* (1951–1957) to Jim Carrey in *The Cable Guy* (1996)—only here used to translate Depression-era America's ideology of "averageness" into comic material. It is Ngai's contention, for example, that zany comedy finds its historic corollary primarily in the experience of "immaterial" labor within the post–World War Two / late-capitalist economy. The way in which information and service sector employees are required to "put affect to work" by performing "friendly reassurance" to customers; how flexible capitalism requires an "absolute adaptability" on the part of its workers, for whom performing a role and doing a job are often one and the same thing; how such a situation moreover places a premium on activity in its own right, irrespective of its material productivity—all of this, Ngai further argues, points to a contemporary "becoming-woman" of post-Fordist employment that takes on qualities paradigmatically associated with domestic work.[124] Zaniness is nothing more nor less than the aesthetic reflex of such a situation.

Yet it is here that the resonance with Benchley's MGM shorts becomes quite inescapable, and it is perhaps no coincidence that Benchley's Joe Doakes everyman is—like the "original" zany, the housewife—primarily a figure of the domestic sphere. Only a handful of times do we ever see Doakes at his workplace—for instance, in the opening minutes of *An Hour for Lunch* (March 1939), *Home Early* (May 1939), and the later *Important Business* (April 1944)—but even then it is only in the context of him *leaving* his office, thereby confirming the general pattern.

Elsewhere, he is shown bringing home a litter of puppies for his family in *How to Train a Dog* (July 1936), spending an evening at home while his wife goes out in *An Evening Alone* (May 1938), caring for an infant in *How to Raise a Baby*, growing vegetables in his backyard in *My Tomato* (December 1943), among other housebound situations. We thus begin to see how, despite Ngai's periodization, zaniness may also have lent itself to a satirical examination of the Depression-era preoccupation with averageness. The zany behaviors of Doakes register a patriarchal perspective on the ways the pursuit of averageness seemingly enmeshed middle-class masculinity in a bewildering set of performances centered upon the private sphere as a site of affective labor.

Perhaps the best translation of these themes in Benchley's filmography comes in *The Day of Rest* (September 1939), whose humor hangs on the conceit of the reversibility of work and rest. In this film, the Joe Doakes character is depicted in his endeavor to enjoy a relaxing Sunday, only to find that relaxation requires a series of performances distinguished from work only by their nonproductive quality. "Working in the so-called garden is another form of so-called relaxation on Sunday," Benchley's narrator wryly informs us at one point; "Moving things up and down from the attic is considered a form of relaxation for Sunday morning," at another. Later, Doakes takes his family out in the automobile for a picnic only to find himself in a traffic jam with "ten thousand other people . . . all headed for the same thing": "The result," we are told, "does not come under the head of relaxation." But it is not only the zaniness of incessant activity that links "relaxation" to the idea of labor, for it further transpires that relaxation is a skill that must be learned, involving above all a sense of timing: one must, in other words, be trained in the labor of efficient relaxation. It is in this sense that the narrator comments at various points that Doakes's "mind has not adapted itself to the idea of relaxation" or that he "doesn't know how to relax," as becomes apparent when Doakes fails to adjust his alarm clock for his weekend lie-in, rises too late to get first dibs on the Sunday paper, or sets out for a picnic at an inopportune time. The conclusion is inevitable: if rest and relaxation turn out to be hard work, then perhaps work is the real respite. As the narrator intones over images of an enervated Joe Doakes struggling to enjoy a game of badminton with his son:

> Just think of that nice cool office where he works during the week. A comfortable swivel chair with an electric fan going. A nice water cooler in the corner. This idea that Sunday is a day for strenuous exercise is undermining the health of our nation. It is tearing down the heart tissues of our manhood. And is probably propaganda started by the fascist or communist nations to make our men unfit for military service in case of war. Besides you look so silly doing it.

If, as Ngai comments, the aesthetic of zaniness "is really an aesthetic about work," then this becomes visible in Benchley's MGM shorts in the cross-coupling of rest and labor in New Deal–era America's pursuit of averageness.[125] And this,

VIDEO 3. Clip from *The Day of Rest* (September 1939).
To watch this video, scan the QR code with your mobile device or visit
DOI: https://doi.org/10.1525/luminos.28.5

ultimately, would seem to be the broader lesson of Benchley's tongue-in-cheek pedagogy, not just in *Day of Rest*: that leisure and labor, two modes of activity that seem separated by a number of sociological divides, are in the final analysis more alike than different (vid. 3).

It is, then, crucial to insist, with Ngai, that the situation in such shorts is not merely one of physical bombardment—of the frustrating effort of moving things around in the attic, for example—but also, and perhaps most centrally, of a kind of affective strain, of the need both to work at relaxation and to put relaxation to work as a performance of suburban "comfort," for this clarifies the distinctions separating Benchley's MGM shorts from a more straightforwardly slapstick mode. To be sure, the Depression-era preoccupation with the average could be and was explored as slapstick. A case in point is RKO's long-lasting *Average Man* series starring Edgar Kennedy (1931–1948), which offers a productive counterpoint to Benchley's MGM work. Both series embodied averageness in the figure of a white suburban patriarch (according to Louella Parsons, author Sinclair Lewis claimed that the RKO series was based on his 1922 novel *Babbitt*).[126] Yet the Kennedy series's very first installment, *Lemon Meringue* (August 1931), signaled the intent to do so in a traditional slapstick vein, culminating in a pie fight. Other initial entries in the series continued to stitch the conventionalized tropes of slapstick into the representation of "average" life and leisure: the second in the series, *Thanks Again* (October 1931), revolves around Kennedy's mishaps with transportation technologies, here an airplane; the third, *Camping Out* (December 1931), involves a standard "hunting trip" slapstick plot, and so forth. Nor was Kennedy alone in this respect: the slapstick depiction of average life was also prominent in other short-subject lines featuring embattled suburban patriarchs—for instance, the long-running Leon Errol series, also at RKO (1934–1951), as well as Columbia's Walter Catlett (1934–1940), Charley Chase (1937–1940), and Hugh Herbert (1943–1952) shorts. But this was not exactly the path of Robert Benchley, whose short subjects for MGM proposed an alternate strategy of comedic representation. Whereas slapstick tends paradigmatically to devolve into what Umberto Eco calls the "comic effect"—that is, a kind of laughter that originates from the viewer's attitude of separation and distance vis-à-vis a hyperbolically physicalized clown—the zaniness of Benchley's humor worked to establish a more equivalential chain binding viewer and viewer's surrogate (Benchley/Doakes) in a shared acknowledgment of quotidian grievances—of the hassles of being unable to find, say, the perfect lamp to read a book (*How to Read* [August 1938]), or the perfect temperature for a shower (*How to Start the Day*), or the perfect seat to watch a movie (*A Night at the Movies* [November 1937]).[127] Neither slapstick clown nor sophisticated absurdist, the Benchley of the mid- to late 1930s established a matrix out of which averageness emerged in comic form as an inclusive term of address and an object of civic identification.

*

Where, though, does this discussion of Benchley leave an understanding of the relation linking comedic forms to the operations of cinematic mass culture? The early parts of this chapter identified two modes of the film industry's address to its public: on the one hand, a rhetoric of difference and distinction, as associated with the marketing of metropolitan sophistication; on the other, a more civic-egalitarian rhetoric of equivalence, associated with the industry's New Deal–era refashioning. But both of these, it can now be seen, were enacted in precisely corresponding ways in the evolving strategies of Benchley's humor—the logic of difference manifest in the differential rhetoric of absurdism, in the cuckoo enumeration of incompatible signifiers; the logic of equivalence in the imputed solidarity of zany averageness in his "How to" shorts. Changes in Benchley's modes of rhetoric and address in this way corresponded to and were coterminous with changes in the industry's imaginary of the mass audience.

Thus did the humorist who had once poured scorn on Hollywood as a place inhabited by "ordinary people" end the decade as one of the nation's preeminent spokespeople of civic ordinariness. Jack Chertok pointed to this trajectory when he described, as we have seen, how Benchley was "becoming the average man to the public . . . rather than just a comic fellow appealing to sophisticated audiences."[128] But Chertok's was, by the late 1930s, just one of many voices to lionize the humorist as the nation's favored interpreter of quotidian annoyance. Despite having "a reputation as being a 'smart' comedian," one commentator noted, Benchley now "mirrors the average man."[129] "One fact is certain," a critic in the *New York Sun* added, "[Benchley] is forever mimicking Mr. and Mrs. John Q. Public," observing that the humorist now "disclaims being a sophisticated wit at all."[130] To be sure, Benchley's "How to" shorts performed their civic pedagogy in motley; still, they did so in a way that acknowledged, rather than merely ridiculed, the idea of a shared commitment to "averageness" through which Depression-era identities were conceived.

Changes in Benchley's comic persona were in this way intertwined with the embattled trajectory of the short subject during the 1930s as object lessons in the variant forms in which the "masses" were imagined by the era's culture industries. To quote one of the founding insights of cultural theory: "There are in fact no masses; there are only ways of seeing people as masses"—ways of seeing, moreover, that became palpable in the short subject's changing modes of address and the forms of comicality it sustained.[131] The development of Benchley's film roles appears, from this perspective, not as a merely personal evolution, but as a reflex of broader film industry endeavors to develop nonslapstick comedic forms pertinent to these different "ways of seeing." Still, as the example of RKO's *Average Man* series reminds us, not all companies were quite so quick to give up the slapstick ghost, and in part 2 of this book we will examine in detail how three producers/distributors of slapstick shorts negotiated the changing position takings that reshaped film comedy during these years. How did these companies adapt their operations to a film industry market in which they were increasingly marginalized? What did the "stigma of slapstick" entail for firms that continued to specialize in the genre? It is to these questions that our analysis now turns.

PART TWO

Case Histories

"The Spice of the Program"

Educational Pictures and the Small-Town Audience

"What the hell's educational about a comedy?" asked slapstick producer Jack White in an interview toward the end of his life. "Something that was very offensive to me," he continued, "was [the slogan] . . . 'This is an Educational Comedy.' There's no such thing as educating yourself with a comedy. It's a stupid name."[1] The object of White's ire? The company for which he had produced and directed two-reel shorts for over a decade—the comedy distributor with the most unlikely of names: Educational Pictures.

The company had been formed in 1915 as the Educational Films Corporation by real-estate man Earle W. Hammons, with the intent indicated by its name: to provide educational subjects for school, church, and other nontheatrical purposes. But by the late 1910s Hammons had realized little profit from this idea and began to target the commercial field, setting in motion a process of expansion that would see Educational become the dominant short-comedy distributor of the late silent era. "It did not take me long to find out that the demand [for educational films] did not exist and that we could not survive by doing that alone," Hammons later recalled.[2] As early as the 1918–1919 season, Educational had begun to diversify its product lines, adding Happy Hooligan and Silk Hat Harry cartoons to its weekly program of travelogues and informational subjects.[3] In April 1920, Hammons signed director Jack White and comedian Lloyd Hamilton from Fox's Sunshine Comedies to produce two-reel comedies under the brand name Mermaid Comedies, and began immediately taking further strides into the comedy market.[4] The program for Educational's 1920–1921 season, which represented the company's first year of general commercial release, included four comedy series: the Mermaids, produced by White; C. L. Chester's animal comedies, featuring "Snooky the Humanzee"; C. C. Burr's Mastodon brand, which produced a series of "Torchy" comedies starring Johnny Hines; and the output of pioneer comedy producer Al Christie. In 1921, Educational picked up for distribution the Punch comedies starring Chester Conklin and Louis Fazenda, among other independently produced series. By the mid-1920s, White's production operations had expanded into what film historian Richard M. Roberts has called a "sort-of General Motors of comedy," offering one- and two-reel product lines to fit all budgets, from the top-of-the-line Mermaids (budgeted at around twenty thousand dollars each) to the mid-range Tuxedo Comedies (around ten thousand dollars each) through to the one-reel Cameo Comedies (five thousand apiece).[5] At its most successful, in 1927, Educational's distribution network extended to some 13,500 theaters (the "widest distribution of any of the [film] companies," Hammons boasted); its output featured two of the era's most noted comedy producers—Jack White and Al Christie, soon to be joined by Mack Sennett, who switched distribution from Pathé to Educational in 1928—along with top-flight comics like Lloyd Hamilton, Lupino Lane, Dorothy Devore, and Larry Semon, as well as the most popular animated star of the 1920s, Pat Sullivan's Felix the Cat.[6] Yet, within a few years of its transition to sound, the company's reputation had sunk. "Educational . . . has released the unfunniest comedies I have ever seen" is one typical exhibitor's report from the mid-1930s. "Another poor comedy from Educational. Why don't they stop making such stuff?" is another. "Educational should have some sort of medal for making the poorest line of shorts of the year," ran a further complaint.[7] To the extent that the company is even acknowledged in film histories today, it is largely as a byword for the perceived wretchedness of short comedies from the early sound era. ("If one searched for a key word to describe the Educational comedies of the 1930s, the best one might be 'cheap,'" wrote Leonard Maltin in his 1972 survey, *The Great Movie Shorts*.)[8]

This chapter seeks to answer several straightforward questions: What happened? How did the most successful independent short-subject distributor of the late silent era flounder so quickly following the shift to sound? In addressing these issues, the chapter seeks not simply to provide an account of the specific missteps and obstacles that undermined Hammons's organization, but also to use that account as a test case for my broader interrogation of the historiographic models that have framed the slapstick short's sound-era decline. By and large, most historians have explained slapstick's changing fortunes during this period in one of two ways: as essentializing aesthetic history (arguing that sound killed the "art" of comic pantomime) or as a kind of social Darwinist industrial history (examining how independent producers of slapstick shorts were squeezed out by the vertically integrated majors). In pursuing my investigation, I want to unpack these models to show how their underlying premises in each case bespeak changing patterns of cultural capital in Depression-era America. The history of Educational Pictures lends itself quite well to this more expansive consideration of historical determinants: a shorts company, it shifts understanding of slapstick's fate away from the individual biographies of the feature-length clown "artists" (away, that is, from the obduracy of a Chaplin, the hubris of a Langdon, the divorce and alcoholism of a Keaton as explanatory factors); an independent, it clarifies the complex adjustments to market conditions necessary to sustain the company's audience against the distribution might of the majors. In both respects, it opens onto a neglected aspect of Depression-era cultural politics whose battleground, we will see, was the very terrain on which "hokum" thrived.

"AN ENTIRELY NEW FORM OF ENTERTAINMENT": EDUCATIONAL AND THE TRANSITION TO SOUND

Perhaps nothing is more established than the perception that slapstick's decline was, first and foremost, a matter of aesthetics—a falling off, as it is often framed, from the beauties of comic pantomime toward the blunt physicality of, say, the Three Stooges. "To put it unkindly"—James Agee wrote in his famous 1949 *Life* essay, "Comedy's Greatest Era"—"the only thing wrong with screen comedy today is that it takes place on a screen which talks."[9] The explanatory framework, we have seen, is one familiar from classical film theory, pitting the putative realism of sound at loggerheads with an idea of art and judging sound an obstacle to the expressive possibilities of comic performance.[10] It is a perception that Charlie Chaplin clearly shared, declaring in 1929 that talkies were "ruining the great beauty of silence" and famously avoiding synchronized dialogue until his 1940 Hitler parody, *The Great Dictator*.[11] And it is a position that would be taken up in subsequent decades by critics like Gerald Mast and Walter Kerr, in language that frequently echoed the insights of film theorist Rudolph Arnheim. Silence, Kerr argued, was "the subtraction [from reality] that guaranteed films would be, so long as they remained mute, flights of fancy"—a premise that cribs from Arnheim to define silent comedy's special artistry as a "fantasy of fact."[12]

The argument that sound killed the art of slapstick has been a hugely prevalent one, and there can be no doubt that comic filmmakers experienced this transition as a challenge of the first order. What can be queried, however, are the terms through which that challenge was experienced and negotiated on the ground, as it were, and it is here that a closer look at Educational Pictures can prove helpful. Amid the great complexity of the company's transition to talking pictures, two facts about the aesthetic implications of sound technology stand out. First, Educational's leading filmmakers were primarily preoccupied not with an idea of comedic *art*—the concern of later critics like Agee and Kerr, as well as pretentious exceptions like Chaplin—but instead with sound's implications for comic pace and tempo. Second, in the case of those exceptions, it was *sound,* not silence, to which the concept of art was most commonly attached—at least, as will be shown, during the initial phase of Educational's transition. In both respects, moreover, these positions took place within the context of Hammons's hesitations and missteps in adapting to sound, and it is here that analysis must begin.

The story of Earle Hammons's initial reaction to sound reads like a stereotype of the industry conservative who failed to see the new technology as anything but a passing fad—at least as Jack White told the tale. "Hammons wouldn't go for sound when everybody else did," White remembered. "I said, 'This will kill us if we don't make talkies right now.' He said, 'It won't kill me. I don't agree with you.' He made a big mistake."[13] Whether or not the characterization is valid is unclear. What is clear is that it was not until January 1928 that Hammons elected to swim with the tide of technological change, by which time most of the major companies had already spent a year in a coordinated investigation of the various sound systems and were on the brink of deciding which of the competing technologies to adapt. Yet it was this exact moment that Hammons unwisely chose to beat the other studios to the punch by gambling on David R. Hochreich's Vocafilm Corporation of America, a sound-on-disc system that the majors had refused even to consider after a disastrous trade debut at New York's Longacre Theater just five months earlier. (The Vocafilm system had at that time been criticized for a "great deal of static" and amplifiers "a bit out of whack with one registering unusually loud and another so faintly it could scarcely be heard").[14] A little over a month later, however, Hammons learned that Paramount, First National, United Artists, Loew's-MGM, and Universal had all decided to sign up for Western Electric's sound-on-film technology. Not wishing to be left out of the pack, he immediately broke his Vocafilm contract and joined the Western Electric contingent.

Such ill-advised wavering ensured that Hammons lost the competitive advantage he had sought in the Vocafilm arrangement and allowed the Warners' Vitaphone shorts to further steal their lead in the changing market. By the time Educational began releasing its first sound shorts—with Mack Sennett's *The Lion's Roar* on December 12, 1928—a number of the vertically integrated majors were also wetting their feet in the field of sound short production (MGM's *Metro Movietone Acts* debuting in September 1928 and Paramount's first sound shorts appearing the following January), while Warners had upped the frequency of its Vitaphone releases to four per week. "It was too little, too late," White recalled. "[Hammons] allowed Warners to make at least 250 talking shorts—musicals, etc.—and when he came along a year later with a talking comedy under his arm, exhibitors said, 'We don't need you.' It hurt him financially. He had a chance a year earlier for us to make sound comedies."[15]

Just as important, these hesitations allowed Vitaphone to define the possibilities and potential of the early sound short, at a time when Educational could only wait on the sidelines. By the 1928–1929 season, as we have seen, Warners was already marketing its shorts in terms of a new and influential reading of distinction premised on a Broadway model of urbane sophistication; few shorts companies, Educational included, remained entirely insulated from these trends as they prepared to make the jump into sound. Yet understandably, Hammons's organization was just as invested in trying to sustain the slapstick comedians and series that

ADVENT OF TALKING PICTURES OPENS NEW FIELD FOR SCREEN

First of Coronet Talking Comedies, "The Eligible Mr. Bangs," Is Last Word in New Screen Art.

(Prepared as advance publicity story)

That the advent of talking pictures has opened up entirely new fields for motion picture producers is evidenced by the announcement of the formation of Coronet Talking Comedies, a Los Angeles corporation, which will produce a series of six comedy playlets using one-act stage material written by well-known writers.

Distribution of the pictures has been secured by Educational Film Exchanges, Inc., and the first of the subjects, "The Eligible Mr. Bangs," is already finished and on the screen. It will be shown at the...........Theatre.........

EDWARD-EVERETT-HORTON·

For Electro Order No. 5732-E

For Mat (Free) Order No. 5732-M

FIGURE 16. Press sheet publicity for *The Eligible Mr. Bangs* (January 1929), the first in Educational's *Coronet Talking Comedy* series, starring Edward Everett Horton. Courtesy Billy Rose Theatre Collection, New York Public Library.

had long been its stock-in-trade. The result, as it played out in studio publicity, was a confusing and contradictory sense of both change *and* continuity regarding the company's first sound releases. On the side of change, Educational's first sound season kowtowed to the new Broadway model by including a new line of six *Coronet Talking Comedy* playlets. Based on stage farces and starring Edward Everett Horton, the series was promoted in ways that asserted the films' theatrical associations, in diametric opposition to the older slapstick credo (fig. 16). "Subtlety, a quality long missing in short comedies, has at last arrived on the screen via the talking picture," announced the exhibitors' press sheet for Coronet's *Prince Gabby* (September 1929), a two-reel comedy about a gentleman burglar, continuing: "Screen comedy for the past two decades has been a thing of fast action, broad situations and physical 'gags.' The new talking picture permits of subtlety of expression through carefully written, clever dialogue and the artistry of the actor in delivering the spoken lines."[16] Indeed, although Horton himself was no stranger to two-reel comedy (having appeared the previous season in a series of starring shorts produced by Harold Lloyd's Hollywood Productions), Coronet publicity chose to emphasize *not* his previous film successes, which went unmentioned, but his theatrical background and experience, describing him as, for instance, a "stage favorite of many up-to-date successes" and a performer "with a successful stage career to his credit."[17] Theatricality was evident, too, in the films' visual design, which, with proscenium-like staging, multiple-camera shooting, and unbroken interior spaces, was seemingly designed to accentuate, rather than mask, the films' stage sources: a representative instance, *Ask Dad* (February 1929)—the second in the series—takes place entirely in a secretary's office, with "action" limited to characters' entrances and exits, and an editing rate sluggish even for the early sound period (average shot length 20.7 seconds, compared to an industry-wide average of 10.8 for the period 1928–1933).[18]

On the side of continuity, however, were those filmmakers and observers who saw these same developments as jeopardizing the formal norms and achievements of silent-era slapstick and struggled to maintain them. The key battleground here emerged around the issue of pace—understandably, given the leaden editing tempo of such virtual theater as the Coronet shorts. Certainly, no other feature of early sound comedy drew as much specific comment from exhibitors, who remained adamant in their complaints about tempo: "The trouble is with action," noted one Idaho exhibitor about recent short comedies. "It is slow, and the stunts are hooked together in a slow, forced manner." "And then there's comedy," lamented another showman discussing recent short features. "Here's where sound has had the most stultifying effect."[19]

Such complaints were hardly limited to comedy. A number of well-known technological difficulties in synchronizing dialogue, prior to 1930, prompted a more or less continual discussion of sound's flattening effect on tempo, regardless of genre: limited mobility for cameras housed in soundproofing blimps, limited actors' movement before the adoption of boom mikes, deliberate and slow dialogue readings to ensure registration—all of these posed problems for what film historian Lea Jacobs calls the "rhythmic control of cinema" across live-action genres.[20] Yet if slapstick remained a special case (where sound had been "most stultifying"), this was because the form had come to be codified during the silent era through a technical convention that sound disallowed: a higher frame-rate for projection. By the 1920s, comedies were typically being projected at a notably faster speed than used during shooting—with a shooting rate ranging anywhere from twelve to twenty frames per second, with variation for effect, and a projection speed of around twenty-two to twenty-four—resulting in an overall buoyancy of comic movement. (Dramatic genres would be projected much closer to the shooting rate, usually at around eighteen to twenty frames per second.) Yet with the coming of sound the demands of synchronization and a stable sound pitch meant that filmic time could no longer be a flexible value; the standardization of motorized cameras and projectors mitigated against the undercranking effects on which silent comedy had depended.[21] It was, then, not only dialogue scenes that flattened pace in slapstick, but the technological apparatus of sound cinema itself.

One can sketch some approaches to these dilemmas through a brief survey of Educational's filmmakers. For few of Hammons's top-producing talents were these issues so pressing as for Jack White, the company's longest-standing producer and a filmmaker with a particular reputation for "fast action" slapstick. The transition to sound not only saw White's return to the director's chair for the first time since 1922—for a series of five *Jack White Talking Comedies*, beginning with the noisily titled *Zip! Boom! Bang!* (March 1929)—but also entailed the challenge of reworking his "fast action" approach within new formal and technological parameters. "Years ago Mr. White introduced a new style in comedy making—'fast action'—meaning that something happened every minute," ran publicity for the series, before reassuring readers that "he continues his fast action in the making of dialogue pictures."[22] On the one hand, this meant restricting the dialogue to bare essentials, an approach that became something of a commonplace in the era's literature on sound tempo.[23] As White put it in a press release at the time, "Fast action . . . has come to mean something entirely different since talking pictures arrived. Where in the silent comedies it meant visually fast action, it now means the . . . fast development of plot and rapidity in establishing situations. This means that dialogue for these comedies must be very carefully edited and pruned of all superfluous words."[24] More distinctive, however, was White's response to the genre-specific problem of frame rates, which, he later claimed, prompted him to don his inventor's cap. "I had an invention that had to do with speeding up or slowing sound," he recalled. "I had an electrical engineer draw the plan up whereby I could change speed without making the sound squeak, affecting only the tempo [of the action]. I thought maybe everybody would use it, but nobody cared for it, so I was out $100 to the lawyer and nothing came of it."[25] Whether or not White ever truly tried to develop such a device is unclear; certainly, the anecdote testifies to the creative strategies through which filmmakers often doggedly sought to bend sound cinema to more familiar comic principles.

A quite different response was to offset the normalization of slapstick pacing through the expressivity of sound. Here, the soundtrack was approached less as a limitation to be transcended than as a new resource to be harnessed to the genre's established stylization of physical action. One sees something of this in Mack Sennett's first season with Hammons, where, much like White, he returned to regular directing duties for the first time in years for a series of *Mack Sennett Talking Comedies.*[26] Press releases from Educational's publicity offices may have emphasized Sennett's excitement at the possibilities of comic dialogue ("Dialogue," he was reported as saying, "opens to the producer of the heretofore 'silent' pictures, the immense field of verbal humor"), but, to judge from available evidence, it was the use of sound *effects* that interested him more, opening up avenues for underscoring the frenzied gags that had long defined his comic style. "Every comedy situation," he insisted, "can be immensely improved by proper sound effects, such as the roar of lions, the rumble of an approaching train or the crash of breaking dishes."[27] Film after film from Sennett's first season was promoted in terms of the capacity of sound effects, not fundamentally to alter the principles of comic cinema but rather to "Enhance the Effectiveness" (as one promotional article put it) of Sennett's knockabout stock-in-trade; "the sound of a starting motor, the crack of a stick over a comic's head, music or the roar of a speeding train" all now produced results "better than any comedy creation that the stage or screen has seen heretofore."[28] Programmatically, Sennett's first sound short, *The Lion's Roar* (working title *Peace and Quiet*), was conceived unrepentantly as a picture about noise. As described in the earliest written draft (and followed more or less closely in the finished film):

> Open up on title: PEACE AND QUIET . . . . . . lap-dissolve to . . . .
>
> > Close up of an old automobile going along a cobblestone street, with one rear tire off and running on the rim, making a terrific rattle . . . . .
> > Lap to shot of a big concrete mixer in noisy action . . . . .
> > Lap to a workman or electric riveter on new building . . . . . .
> > Lap to a general shot of busy city street, with usual noises—street car bells, auto horns, newsboys shouting papers, etc. from which—
> > Lap dissolve to . . . .
> > INTERIOR: CLARENCE'S ROOM IN CITY (DAY).[29]

For the rest of the film, Clarence (Johnny Burke) flees the bedlam of urban life to spend a weekend in the country with his beloved (Daphne Pollard), only for his "peace and quiet" to be shattered when he finds himself trapped up a tree during a hunting trip, perched above a bellowing mountain lion. The gag here, of course, is that the country is ultimately no less free of din and disturbance than the city, but at a deeper level, Sennett was simply using sound to cock the same snook that he had been pulling for close to twenty years, creating a carnival of aural cacophony as a straightforward functional equivalent for his trademark visual chaos (fig. 17).[30]

Needless to say, Sennett was hardly alone in appropriating sound effects to established knockabout procedures. It was, in fact, the increasingly widespread use of such effects that provoked Harold Lloyd, who had just completed production on the silent feature *Welcome Danger* (1929), to reshoot the entire film for sound. As he recalled, "Sound was just coming in, and inconsequential things were getting tremendous laughs—like frying eggs and ice tinkling in a glass. They'd howl at that. So I said, here we're working our heads off trying to get funny ideas, and they're getting them from these sound effects. I said that maybe we had missed the boat and should make *Welcome Danger* over."[31] The foundation for such an effects-laden approach had, in fact, already been firmly established by silent-era musical practice, when various noise-making devices—called "traps"—were commonly used in film accompaniment, especially in comedies. Originating in live performances like vaudeville, the "trap drummer" had been responsible for supplying sound effects in sync with the onscreen comic action throughout the silent period, using an assortment of noisemakers for this purpose, from simple coconut shells to more baroque devices.[32] (A cue sheet compiled for Sennett's silent short *Smith's Modiste Shop* [December 1927] suggests just how elaborate such effects could be, including cues for the sound of a boy's slingshot, a smashed ink bottle, and a meat chopping machine.)[33] The practice also—at least by the 1920s—had created significant distinctions according to cultural value, whereby "low" cinematic genres like comedy or animation were permitted a kind of anti-illusionist, nonrealist sound accompaniment (e.g., a slide whistle to accompany a slip) that would have been considered

FIGURE 17. Johnny Burke (with gun) and Billy Bevan, finding no peace in the countryside in Mack Sennett's first sound short, *The Lion's Roar* (December 1928). Courtesy Academy of Motion Picture Arts and Sciences.

the height of vulgarity in "serious" drama.[34] Thus, whereas Sennett's early sound comedies built their soundtracks almost entirely out of *realist* diegetic effects ("the roar of lions . . . the crash of breaking dishes"), the idea of integrating more illusion-destroying "trap"-style noises was a predictable next step—an approach most notably perfected a few years later, not at Educational but at Columbia's short-subjects division, reorganized under the supervision of Jack White's brother Jules in 1933.[35] Under the stewardship of sound effects man Joe Henrie, Columbia's shorts developed an elaborate grammar of knockabout clamor—nowhere more effective than in the shorts of the Three Stooges—translating the quick-paced stylization of silent-era comic action into violent sonic outbursts: face slaps accentuated by the crack of a whip; eye poking by two plunks of a ukulele; ear twisting by the turning of a ratchet; head bonking by a wooden tempo block; blows to the stomach by the sound of a kettledrum, all in quick succession.[36] (Citing just such effects, Jules White would later claim that it was the early sound period, not the silent era, that was the "Golden Age" of slapstick comedy.)[37]

Still, it is at Educational that we see these adjustments occurring in real time—in tandem with the introduction of the technology and not a few years later, as at

Columbia—and the process sheds an often surprising light on how the company's filmmakers used sound less to pursue new directions than to sustain and elaborate upon the old. For all the bluster with which producers like Jack White and Sennett spoke of sound as ushering in "an entirely new form of entertainment" (the quote is White's), the reality was that the new technology was more typically called upon to shore up the established premises of slapstick filmmaking, and not only in the realm of tempo and trap drumming.[38] Such is the case, for instance, in Sennett's third sound film, *The Old Barn* (February 1929), in which sound prompted a surprise return to the procedures of the comic melodrama. A founding staple of Sennett's comic philosophy, the burlesque melodrama had flourished during the filmmaker's late Biograph / early Keystone years as a way of tweaking the moral terms of Griffithian melodrama, blending the thrills of D. W. Griffith's trademark race-to-the-rescue finales with the comic effect of parodic imitation.[39] Yet, whereas the silent burlesques had typically generated humor from characters' misreading of *visual* signs—for instance, Mabel Normand mistakenly believing a curtain's chance movement to have been caused by a burglar's hand in Sennett's Biograph short *Help! Help!* (April 1912)[40]—the comic plot of *The Old Barn* supplements this with a series of misheard *aural* cues. In the film's climactic nighttime sequence, star Johnny Burke leads the guests of a rural boarding house to an old barn to search for an escaped convict, resulting in a series of sound gags in which the would-be detectives mishear an old car's wheezing horn as a man's groan, a burst balloon as a gunshot, and so forth. The same principle was revisited later in 1929, in *The Constabule* (August), in which small-town constable Harry Gribbon and rail agent Andy Clyde spend the night at the station to protect a shipment of money. In a protracted comic sequence—the scripting of which involved Sennett himself—the two characters come to suspect falsely that they are under attack from burglars, first when they see a curtain moving (actually caused by a kitten), subsequently when they hear the explosion of gunshot cartridges (accidentally dropped into a lit stove).[41] As is typical of Sennett's earlier burlesques, the sequence ends with the unmasking of the error and the protagonists' embarrassment: a group of train passengers enters the station building to see the commotion, Gribbon pulls back the curtain to expose the "burglar," and a harmless kitten is revealed. "Kitten, you're under arrest," one passenger snidely remarks.[42]

*

It is evident, then, that, despite sound's unmistakable impact on tempo and pacing, the new technology *could* also function within—and even elaborate upon—canonized tropes and comic formulas. But, if this is the case, then it is equally evident that slapstick's "decline" in the early sound era needs to be understood in terms that go beyond questions of comedic form. To a greater degree than has often been thought, sound *was* assimilable to the formal norms of slapstick film, and in that sense, what changed must also be sought outside the properties of the comic texts themselves. Similarly, it will not do to see the various approaches discussed above solely as a matter of formal continuity versus change, the former represented in Sennett and White's endeavors to harness sound to established slapstick technique, the latter by the "new style" of playlet comedy, exemplified at Educational by the Coronet films. Such a characterization risks obscuring the degree to which broader discourses of taste and cultural politics were also implicated in that division: change, at Educational, meant adapting short comedy to the format of theatrical farce, while continuity implied keeping faith with the knockabout credo of "fast action, broad situations and physical 'gags.'" The coming of sound can thus be seen to have marked an intensification in long-standing divisions separating traditions of sophisticated humor from the sensationalism of "low" comedy traditions, divisions that, with the transition to sound, came to be associated with alternative uses for the new technology: sophistication was equated with the refinements of dialogue, slapstick with the immediacy of sound effects, with the noisy impact of misfiring gun cartridges as well as kettledrum bonks and the like.[43]

It is moreover possible at this point to see more precisely how the later aesthetic readings of slapstick's decline offered by Agee and Kerr go wrong. For they radically misperceive how the idea of "art" was located within contexts of production and reception at the time. The coming of sound was not primarily experienced as a shift *away* from comic artistry—away, that is, from the formal beauties of silent pantomime, as critics like Kerr would later argue. If anything, as publicity for the Coronet series makes clear, sound could be and was promoted as enabling a shift *toward* art, toward the "artistry" of "clever, subtle comedy" as exemplified by dialogue humor in theatrical-style farce. Correspondingly, what was feared to have been lost with sound was not "art," but its opposite—that is, the broad, popular style of slapstick in which producers like White and Sennett had formerly specialized and which they sought valiantly to sustain in the new era. Exhibitors who lamented the impact of sound thus typically spoke not of artistry, nor of the decline of pantomime, but more straightforwardly of the need for a return to "good old-fashioned" or "dandy old-fashioned slapstick."[44] The appropriate dichotomy for comprehending sound's impact on short-format comedy—at least as it was experienced by filmmakers and audiences at the time—thus has very little to do with the Arnheimian division of art versus realism; rather it is the hierarchical gulf that Gilbert Seldes identified in a 1932 essay in which he divided America's comic sensibility into sophisticated and urbane versus populist and provincial modes, a division and an essay to which this chapter will be returning.[45]

"OUR PRODUCT WAS BLOCKED": EDUCATIONAL AND THE SHORT-SUBJECT MARKET

Such, then, are the difficulties that emerge from a brief meditation on formal readings of slapstick's decline, but similar themes soon surface, rearranged in a somewhat different framework, from our second explanatory model, industrial history. In his economic study of the Hal Roach Studios, Richard Lewis Ward has shown how the move by the major studios into the production and distribution of short subjects during the mid-1920s, led by Paramount and MGM, effectively shut independently produced shorts out of the major-owned houses, consigning them to the less profitable terrain of nonaffiliated chains and small-town exhibitors.[46] Independent short companies survived these new economic realities only to the extent that they entered into alliances with the majors. In 1926, for instance, Hal Roach jumped at the opportunity to affiliate with Loew's-MGM and, in consequence, prospered during the 1930s. Mack Sennett, meanwhile, was less successful. Rumors that he was to enter into a new combination with Paramount had circulated since the mid-1920s but only came to pass several years later, in 1932, when Sennett jumped ship from Educational to produce a series of Paramount comedies. When Paramount declined to renew that arrangement for a second season, the Sennett studio immediately floundered, unable to find a new distributor; within months, it was declared bankrupt in federal court in Los Angeles.

Ward's analysis provides a crucial entry point for any assessment of the fate of the sound short during this period; yet, inasmuch as his focus is on the Hal Roach Studios, it understandably offers less detail on the fortunes of the independent companies that, lacking the lifejacket of major studio backing, faced a harder struggle to stay afloat within the turbulent exhibition market of Depression-era America. Initially, Educational might have seemed to be in a fairly secure position; the company was, after all, primarily organized as a distributor rather than a producing concern and, at its most successful in the mid-1920s, had developed its own network of some 13,500 theaters for its films, including contracts with major chains like West Coast Theatres, Stanley, Loew's, and biggest of all, Paramount (from which Hammons estimated his company received "one-tenth of our gross").[47] Yet that network soon proved unreliable. As early as 1927, Hammons was publicly voicing his fears that, following their shift into short-subject production, the majors would now block Educational product from their theaters. "Paramount-Famous-Lasky are going into the releasing of short subjects," Hammons explained to a class of Harvard business graduates in a series of film industry lectures organized by Joseph P. Kennedy. "It is only natural to expect that their theatre department will book all their short reels. These theatres have been a source of large revenue for our company, and we are confronted with the problem of retaining that revenue."[48] Jack White later remembered how quickly such fears were realized: "When Metro started production of shorts and comedies, all the other majors followed suit," he explained. "Our product was blocked. . . . So even though we had captured the comedy market and, by the exhibitors' own admission, had saved their shows time and time again, they had to play the majors' shorts in order to get their features."[49]

One obvious tactic was for Educational to respond in kind, encroaching on the majors' bailiwick—that is, feature-length films—even as the majors were advancing into shorts. In the fall of 1928, Educational did just that, purchasing a 50 percent interest in World Wide Pictures, a new feature distributor formed with the intent to deal "exclusively in films produced in countries other than the US."[50] Hammons next expanded his feature interests by brokering the amalgamation of World Wide with Sono-Art Productions, an independent feature producer, in 1929. Then, in the spring of 1931, he brought another feature company into the mix, this time folding L. A. Young's Tiffany Pictures into the Sono Art–World Wide combine. While so much expansion may not have been wise at the height of the Depression, it did result in a brief spike in investor confidence, as stock in the company jumped almost twenty points following the Tiffany merger.[51] It also saw Mack Sennett's return to feature-length filmmaking, for what would be the final time in his career. Scarcely was the ink dry on the Tiffany deal than it was announced that Mack Sennett would "produce and direct" a feature starring the blackface comedy team Moran and Mack (also called the Two Black Crows)—a move that may have been designed to placate Sennett, who was already considering leaving the Educational fold.[52] That film eventually materialized as the eight-reel *Hypnotized,* released under the World Wide banner to favorable reviews during Christmas week of 1932. Yet it was already clear by this point that Hammons's involvement in features was not working, as World Wide was proving unable to fulfill its exhibitors' contracts. Soon trade press articles were reporting that "exhibitors refus[ed] to play Educational shorts because the company is . . . not releasing the full quota of World Wide features," and Hammons eventually cut ties with the company, which limped on through the 1930s by distributing a dwindling number of imported features through states' rights (including some classics like Jean Renoir's 1937 *La Grande Illusion*).[53]

If Hammons's organization was to succeed in the business circumstances of the early sound era, then it would have to be on the strengths of its short subjects alone and their competitive appeal to exhibitors neither affiliated nor contracted with the majors. Yet, even here, the economic circumstances of the Depression provided a further turn of the screw, as independent theater owners now began embracing the policy of double billing in an attempt to boost attendance. It could hardly have been a surprise that Hammons would emerge as one of the leading voices in the battle against double bills, time and again using the trade press as a bully pulpit to denounce the practice ("The greatest evil the industry has ever known," "an insidious evil," "demoralizing . . . [for] our industry" are some of his quoted opinions).[54] Hammons also moved quickly to take publicity steps seemingly designed with the loyalty of smaller exhibitors in mind. One of these—a personal tour of the exhibition situation in the Midwest in 1932—left Hammons optimistically forecasting "a steady increase in grosses," although no evidence survives to indicate exactly what cities and towns he visited.[55] Also oriented toward the small-town theater owner was a series of Educational advertisements appearing in the trade press that spring. "It Sounds Like a Bargain Once," the ads admitted, but the more sinister truth, they implied, was that double bills were insidiously undermining family values.

One showed a cartoon family of moviegoers bored and angry at having to tolerate two features on a single bill; another implied that double features were the choice only of sinister-looking bachelors; yet another quoted a Mrs. Eunice McClure, of the Illinois Federation of Women's Clubs, claiming that double bills were responsible for making "children too weary to tell [their parents] what they have seen" and were keeping "entire families . . . away from the theatre" (fig. 18).[56]

Such advertisements arguably represent the short-subject industry's most vociferous effort to exploit the rhetoric of consumer protection in opposition to duals—discussed in the previous chapter—and they did so by directly appealing to smaller exhibitors' dependence on the family trade. Unlike metropolitan theaters, which could profit by targeting specific demographics, small-town and neighborhood theaters had to attract the entire potential audience for each film; the way to do that, at least according to these ads, was to ensure "the variety that children and adults demand" by screening "a program of one good feature and several of Educational's short subjects."[57] (Educational in fact repeated this publicity strategy in the summer of 1937, with a series of four ads blaming an epidemic of "doubleitis" for everything from job absenteeism among family breadwinners to housewives' refusal to make breakfast.)[58]

Hammons was also a pacesetter in advocating for diversified appeal as a front against the lure of double bills. As we have seen, the diversification of short subjects encompassed a range of intertwined motives during the early sound period: if it was initially a means for exhibitors to sustain standards of variety in the growing absence of live acts, then within a few years it had become a tactic for short-subject companies to stave off competition from duals.[59] Accordingly, in 1932, Hammons began explicitly promoting diversified programs as a way to combat "ruinous" duals, announcing for the 1932–1933 season "a program of short subjects offering . . . a greater variety of subject matter than ever before in [the company's] history."[60] A comparison with the firm's output in the two preceding seasons shows that this was not mere rhetoric. For 1930–1931, for instance, Educational's eleven series had included only three that were not live-action comedies: the animated *Terry-Toons* and *Hodge-Podge* series and the *William J. Burns Detective Mysteries*, all one-reelers. Two seasons later, the total number of series had expanded to nineteen, of which more than half were not live-action comedies: the two one-reel animated series—*Terry-Toons* and *Hodge-Podge*; two new musical series—the two-reel *Kendall de Vally Operalogues* ("World famous operas brought to the screen in tabloid form"), and Reinald Werrenrath's one-reel *Spirit of the Campus* films ("Showing the life and spirit of our famous universities, with their songs sung by the noted opera and radio baritone"); three scenic and educational series—*Camera Adventures*, *Bray's Naturgraphs*, and *Battle for Life*; the one-reel entertainment

THE PUBLIC'S PREFERENCE IS CLEAR

VOTE HERE FOR
TWO LONG FEATURES

THERE IS NO CHANCE OF MISTAKING
THE VERDICT . . . .

WHEREVER PUBLIC OPINION IS
RECORDED, IT'S A LANDSLIDE

VOTE HERE FOR
ONE BIG FEATURE
EDUCATIONAL COMEDY
NEWSREEL
NOVELTIES

NEWSPAPER READERS, CLUB MEMBERS,
THEATRE PATRONS, ALL GIVE . . .

A TREMENDOUS MAJORITY IN FAVOR
OF THE DIVERSIFIED PROGRAM

More showmen are applying the "give the public what it wants" rule to the double feature question. More patrons are being asked to express their opinions. And everywhere the balloting is a positive demand for diversified entertainment. You'll keep your audience—and yourself —happier if you meet the public's wishes with a good variety show...one fine feature, one of *Educational's* rip-roaring two-reel comedies and plenty of novelties selected from *Educational's* one-reel pictures. Try it now and prove for yourself that it's "what the public wants."

EDUCATIONAL FILM EXCHANGES, *Inc.*

Educational Pictures
"THE SPICE OF THE PROGRAM"

E·W·HAMMONS
President

FIGURE 18. One of Educational's anti-double-bill advertisements, from *Motion Picture Herald*, April 30, 1932.

newsreel *Broadway Gossip;* two series recycling silent film footage with ironic commentary from humorists Harry Miller and Lew Lehr—the serial satire, *The Great Hokum Mystery* ("It was once a thr-r-r-illing drama, but now . . . it is a comedy riot"), and the nostalgia series, *Do You Remember?* ("Memories of the Gay Nineties at their gayest. With a line of chatter by Lew Lehr and Harry Miller that will keep any audience in an uproar of laughter"); and the two-reel *Gleason's Sport Featurettes*, a short-lived attempt to fuse the format of the sports short with comedy narratives. Moreover, of the eight live-action comic series on the 1932–1933 program, at least one departed substantially from conventional slapstick—the one-reel *Baby Burlesks* ("Satires on the big screen hits, enacted by tiny tots," starring a four-year-old Shirley Temple in what was her screen debut).[61]

Yet even this bid for diversity failed to provide a toehold for the company, saddled as it was with a number of underperforming assets. Not only was the World Wide feature slate creating widespread headaches for exhibitors, but the Christie Film Company (which had returned to Educational in 1931 after three seasons with Paramount) was providing an additional drag on profitability, having unwisely invested much of its assets in the sound conversion of the faltering Metropolitan Studios. The last of the large independent short producers still outside the majors' control, Hammons's organization was finally forced to relinquish its independence early in 1933, when Educational's creditors, led by Chase Bank and Electrical Research Products Inc. (ERPI), stepped in to restructure the company. Educational was required to close all of its exchanges (a reduction to overhead of some twenty thousand dollars per week) and immediately entered into a life-saving merger with the Fox Film Corporation.[62] The company's product would now be distributed through Fox's distribution network, while Fox took over Educational's existing exhibition contracts.[63]

One positive consequence of the new arrangement, according to the trade press, was "a material rise of bookings [of Educational films] into first-runs in New York," as Educational films now had access to Fox-affiliated theaters.[64] Crucially, though, the deal with Fox did nothing toward solving the double-feature "evil," and Educational's fortunes continued to trend downward as the practice spread. In 1934, Hammons began closing up the company's Hollywood studios, first relocating about half of Educational's production operations to its Eastern Service Studio in Astoria, then completing the move two years later. Even the majors were, by this point, caving in to the pressure of competition from double features, implementing "B" unit production strategies on studio lots and allowing some of their first-run affiliated houses to screen double bills. The lifesaver cast by the deal with Fox quickly turned into a stone as a number of the major studios, led by Paramount, now sought to cut two-reelers from their product lines.[65] "Double features ruined [Hammons]," Jack White later recalled. "[Exhibitors] wouldn't tolerate him anymore, and they didn't have to because they had double features. They could afford to tell Mr. Hammons and his product to shove it."[66] The company finally lost its uphill battle in 1938 when Twentieth Century–Fox cut ties with Educational, declaring "no market for two-reel shorts because of dual bills."[67] There was one final misstep, when Hammons merged all of Educational's assets with the failing Grand National Studios in another ill-advised bid for the feature market. By 1940, swamped in debt, Grand National was liquidated, and Hammons's quarter-century involvement in the commercial film industry ended.

"IT'S OLD STUFF BUT IT MADE THE FARMERS LAUGH": EDUCATIONAL AND THE SMALL-TOWN AUDIENCE

What begins to come into focus at this point is a significant structural homology linking the two trajectories of analysis thus far. For both the conventional aesthetic interpretation (sound killed the "beauties" of pantomime) and the industrial explanation (short-subject producers succeeded only through allegiances with the majors) can be seen to open onto a series of dichotomies splitting the field of short-format comedy in the early sound era. The aesthetic split juxtaposing Broadway-style sophistication against popular standards of "fast action" slapstick, the divided market pitting the metropolitan first-run circuits against second-run and small-town chains—these are related dichotomies that speak to much broader cultural divisions during this period. And it is in this sense that the fate of the early sound short finds a further horizon of interpretation within changes in the very structure of Depression-era mass culture.

One needs to return here to the growing distance separating small-town from urban moviegoing cultures, touched on in earlier chapters. In part a function of widening disparities in the context of an urbanizing nation, urban-rural tensions suffused American life in this period. Rural and small-town people had predominated within the nation's identity at the end of the previous century (they still represented 70 percent of the nation in 1900), but had dropped to under half of the population by the 1930s—a decline that, combined with the social dislocations wrought by the Depression, intensified anxieties about the place of small-town and rural values in the mainstream of US culture. Indeed, as James Shortridge has argued in *The Middle West: Its Meaning in American Culture,*the prestige of rural states within the nation's imaginary had already been seriously eroded from around 1920, a year that saw not only the beginning of a major agricultural recession but also the publication of Sinclair Lewis's biting satire of small-town Minnesotan life, *Main Street.*[68] Yet, as pastoral ideals fell increasingly out of step with a modernizing society, rural residents nonetheless struggled to stake out a cultural identity as more than just "those who stayed behind."[69] A new assertiveness was expressed in various forms of "regionalism" across the political spectrum, encompassing anything from the white supremacist nostalgia of the southern Agrarians to the emergence of a "new regionalism in American literature"—to quote California writer Carey McWilliams—in small magazines like *Folk-Say* and *Space,* whose content often overlapped with the proletarian avant-garde.[70] As Michael Denning has noted, the appearance of inclusiveness sought by New Deal–era populist rhetoric was thus betrayed by deeper structural divisions, of which regional grassroots movements were a significant symptom.[71]

Within the film exhibition market, meanwhile, small-town exhibitors became increasingly vocal in their complaints about Hollywood's trade practices as they struggled to differentiate themselves and their publics within the larger cultural field. The sense of marginalization was sharpened, in the first place, by the palpable imbalance of power wrought by the transition to talking pictures. The expense of sound installation had been overwhelming for small independent exhibitors—in 1929, ERPI charged seven thousand dollars to wire theaters with five hundred seats or less—leading many to sacrifice local autonomy by selling out to larger, city-based chains, if they did not simply shutter their doors.[72] Those that struggled through were then further hit by the economic downturn, which saw box-office revenues fall off by a third. The small-town theatrical market was decimated: metropolitan centers on the coasts were fortunate to experience closure rates of between 7 and 20 percent, but the Midwest, the South, the Plains, and northern New England lost anywhere from 22 to 48 percent of theaters.[73] In such a context, columns like *Motion Picture Herald*'s "What the Picture Did for Me"—a forum for exhibitors' comments on current films—became a lifesaver for nonmetropolitan theater owners, a sounding board for demands for the production of films that would suit their box-office needs.[74] Economic marginalization thus played into emerging divisions of taste, as smaller exhibitors now began to forcefully complain about movies' urban bias. As J. C. Jenkins, the *Herald*'s regional correspondent, complained in a 1933 article, "Smutty dialogue and nasty suggestions, illicit love scenes and the like may get a 'kick' from city audiences but they are kickbacks from rural communities."[75] Whereas the initial conversion period saw film producers promoting an imaginary continuity linking small-town moviegoing to metropolitan cultural centers, the early to mid-1930s saw growing regional resistance to such strategies, as local exhibitors defiantly asserted local values, calling for films that would better suit small-town needs—action and adventure films, comedies, musicals, and "American" characters.

Of course, the rhetoric of cultural division cut both ways. Already by the 1920s, a whole new vocabulary of distinction was coming into use that disparaged rural America for the perceived naïveté and simplicity of its cultural tastes, foremost among which, indeed, was "hokum" (or "hoke"). In addition to its other connotations—discussed in my introduction—"hokum" thus crucially served during this period to crystallize many of the assumptions about the preferences of small-town audiences, in particular their supposed fondness for strong effects and overt moralizing.[76] "We—Want—Hokum!" proclaimed the title of fan magazine *Picture Play*'s

1927 exposé of the tastes of rural moviegoers, continuing: "Does the average fan really like all these big, supposedly artistic films that are being made for him nowadays, or wouldn't he much rather see a good old-fashioned melo-thriller, slapstick comedy, or rip-roaring Western film?"[77] Hokum, in this sense, implied a kind of cultural anti-modernism—a taste for "old-time" or "good old-fashioned" entertainment—and the term became a pivot around which emerging cultural divisions took shape. In the hands of *Variety*'s urbancentric writers, the word was commonly meant as a term of denigration, where the taste for "hokum" implied a kind of hayseed backwardness; yet the word was also mobilized as a badge of honor for small-town publics who resisted the suspect sophistication of metropolitan cultures—as when one Kansas exhibitor evoked the superiority of "custard pie hokum" as the "real stuff" in comparison to pretentious "Pulitzer prize plays."[78] Hokum, in short, designated the way geography became cultural capital, expressed through aesthetic distinctions and presumptions regarding audience dispositions and tastes.

As the above reference to "custard pie hokum" suggests, moreover, comedy played a key role within this process of cultural position taking. The small-town market had long been considered a reliable one for slapstick producers (as early as 1924, Mack Sennett had spoken of the small-town audience as the "real acid test" for slapstick producers), but, as we have seen in earlier chapters, the perception of an alignment between popular humor and hinterland tastes had greatly intensified by the Depression's earliest years.[79] Gilbert Seldes, in his aforementioned 1932 essay, *defined* popular humor as that which is "specifically adapted to the small town citizen, the rustic, and the provincial," while Constance Rourke's landmark cultural history, *American Humor* (1931), reinterpreted the entire tradition of US literary humor from the perspective of localism ("the very base of the comic in America," in Rourke's assessment).[80] Long-established hierarchies separating "low" comedy from "sophisticated" humor—distinctions that, earlier in the century, had been coded primarily in terms of class difference—were increasingly recast in relation to the small-town/metropolitan split orchestrating Depression-era mass culture. One of the earliest sociological studies of rural audiences—"Rural Preferences in Motion Pictures," published in a 1930 *Journal of Social Psychology* by Harold Ellis Jones and Herbert S. Conrad—corroborated the general perspective, albeit by making a somewhat unscientific appeal to general observation: "An *observational* study of the responses of rural and urban audiences to comedy reel episodes shows, in the former group, a franker and more boisterous delight for the slapstick types of situation," a preference they baldly attributed to the "psychological crudities" of rural audiences.[81] The issue, then, for independent short companies like Educational was not only that the 1930s saw a weakening of the slapstick short's industrial position; it was also the more complex process that had witnessed an emerging split in the nation's exhibition market and a corollary change in the cultural affiliations of knockabout comedy, its increasing marginalization as small-town "hokum" within the cultural hierarchies of the period.

Of course, slapstick was hardly the only cinematic genre to take shape within these emerging taste hierarchies. Peter Stanfield's study of the 1930s western has shown how "B" westerns were conceived and organized around the assumption of a small-town audience, as evident in "singing cowboy" films that exploited the fad for hillbilly and cowboy songs.[82] A similar situation had previously applied in radio, which, as early as the mid-1920s, had targeted the rural market with "barn-dance" musical extravaganzas, such as Nashville's *WSM Barn Dance* (renamed, in 1927, the *Grand Ole Opry*) and Chicago's *National Barn Dance,* to say nothing of the astonishing number of radio comedies with country store settings (*Lum and Abner, Eb and Zeb, Si and Elmer, Ike and Eli, Lem and Martha, Herb and Hank, Rufie and Goofie,* etc.).[83] In all these cases, country music and comic rube characters served as a primary means by which manufacturers and radio sponsors pitched their product to the rural working class, offering regional listeners a sense of identity and community against the traumas of dislocation, disenfranchisement, and dispossession brought on by the Depression.

Educational, too, played a similar game, particularly in the early 1930s, when its product was still largely frozen out of the metropolitan, major-owned circuits. One sees this, for instance, in the shifts within the musical short series that Educational first introduced in the 1932–1933 season: whereas the earliest of these, the *Kendall de Vally Operalogues,* had evidently gestured toward older ideals of highbrow culture—and, perhaps in consequence, had been judged "no good for the small town" by one Idaho exhibitor—Hammons's organization soon began adding more regional forms of appeal, most notably with its *Song Hit Stories* and *Song and Comedy Hits* lines, produced by Al Christie at Educational's Eastern Service Studios in Astoria.[84] The most enduring and consistently popular of Educational's musical series (lasting from the 1933–1934 season until 1937–1938), the *Song Hit Stories* and *Song and Comedy Hits* were song-filled sketches running the gamut of musical styles—from seafaring ballads in *The Bounding Main* (November 1934) to gay nineties nostalgia in *Gay Old Days* (January 1935)—but with a particular emphasis on the rural vernacular. Styles like country and western (with the ubiquitous "Home on the Range" popping up in western-themed shorts like *The Last Dogie* [November 1933] and *Rodeo Day* [September 1935]), hillbilly (in shorts like *Mountain Melody* [August 1934] and *Hillbilly Love* [October 1935], the latter featuring Frank Luther from the NBC radio series *Hillbilly Heart-Throbs*), and southern black music (with Stepin Fetchit and Lethia Hill in *Slow Poke* [September 1933] as well as numerous shorts featuring the Cabin Kids)—all contributed to the series' consistent acclaim in the *Herald*'s "What the Picture Did for Me" column, where one Missouri exhibitor spoke of them as "the best of the single reels by Educational."[85]

The question, then, becomes whether the imprint of the small-town market can also be traced in Educational's *slapstick* output. Certainly, approaching Educational's product from this perspective clarifies a number of otherwise perplexing developments, in particular the surprising stardom of two frequently paired comedians

FIGURE 19. Press sheet publicity for *The Constabule* (August 1929), a *Mack Sennett Talking Comedy*, with Andy Clyde (left) and Harry Gribbon. Courtesy Billy Rose Theatre Collection, New York Public Library.

who dominated Mack Sennett's early talkie output at Educational: Andy Clyde and Harry Gribbon (fig. 19). Often cited as evidence of the tough times on which the Sennett brand had fallen, the unlikely ascendancy of these two comics might profitably be read as a revealing barometer of slapstick's shifting cultural valences. The period surrounding the coming of sound, it should be noted, had represented a significant reshuffling within the upper echelons of Educational's comic talent. Lloyd Hamilton, the company's biggest star, had been barred from the screen by the MPPDA for the 1928–29 season, following a series of arrests for public drunkenness and an Arbuckle-style scandal in which the comedian's name had been brought up in association with a nightclub shooting.[86] Next Lupino Lane—Educational's second-biggest name and its most highly promoted comedian in Hamilton's absence—departed the company in 1929, eventually quitting Hollywood altogether to return to his native England the following year. Nowhere, though, did this changing of the guard produce more telling consequences than on the Sennett lot, facilitating a shift toward rural characterizations and settings in the studio's early sound output. When Sennett had begun his distribution arrangement with Educational, his leading comedian was the former Broadway performer and big-time vaudevillian, Johnny Burke, who had risen to fame on the stage for his "doughboy" routine and who joined Sennett in December 1926.

Yet Burke's relationship with the Sennett studio barely survived the transition to sound following a pay dispute. (He was the highest-paid comedian on the lot—with a weekly salary of sixteen hundred dollars by early 1929—and Sennett was infamously tight-fisted.)[87] Burke's departure subsequently cleared the way for Clyde and Gribbon, who had first appeared together as rural characters in the Burke vehicle, *The Bride's Relations* (January 1929)—playing Johnny's hick in-laws, Clyde as Uncle Ed, a "jovial type farmer," and Gribbon as the outsized simpleton Homer—and whose subsequent pairings would dominate Sennett's first year of sound production.[88]

A one-time stage actor who had starred in musical comedies produced by George M. Cohan, "Silk-Hat" Harry Gribbon had in fact been a veteran of Sennett's studio from the Keystone days, initially signed in 1915 as part of an effort to hire "high-hat"-style comedians with genteel appeal. Although the subsequent decade had seen him move into features (as well as touring in vaudeville), Gribbon had reunited with Sennett for the Educational films, where, in a striking reversal of his earlier persona, he was now marketed as a specialist in small-town "boob" roles—frequently in the same "Homer" characterization first assayed in *The Bride's Relations*. Ditto Andy Clyde: a Sennett regular throughout the 1920s, Clyde's persona underwent a similar reevaluation with sound, emerging in the persona of "Ed Martin," a countrified old-man type—despite Clyde's actually being in his thirties—directly in the tradition of the cracker-barrel patriarchs then popular on radio. (The producer Jules White, for whom Clyde would subsequently work at Columbia, referred to him explicitly as a "hick" comedian.)[89]

Gribbon and Clyde's first top-billed pairing was *Whirls and Girls,* released in February 1929, and they subsequently starred in no less than ten of fifteen Sennett releases over the subsequent twelve months, most commonly in films with rural themes and settings like *The Big Palooka* (May), *The Constabule,* and *The Lunkhead* (September).[90] As early as the second of these films, *The Bees' Buzz* (April), Sennett's scenarists—of whom the core team for these films was Harry McCoy, Earle Rodney, and Hampton Del Ruth—had established the basic story formula that would provide the series' framework. Clyde's "Ed Martin" character is father to an independent young woman, played by Thelma Hill, who in turn is the object of Gribbon's bumbling affections. Thelma, however, favors another, typically a "straight" juvenile lead who contrasts with Gribbon's rube-ish clown. The films thus operate within the logic of a comic love triangle, pitting Gribbon's rural idiocy against the decidedly nonrural—hence more "normalized"—traits of the rival suitor: for instance, Gribbon's "village boy" Homer versus Thelma's college sweetheart (Milton Holmes) in *The Constabule;* Gribbon's rube-in-the-big-city, Gilbert, versus another college sweetheart (Ben Alexander) in *The Lunkhead;* or Gribbon's vulgar western oil tycoon, George Palooka, versus champion California golfer Charlie Guest (playing himself) in *The Golfers* (September 1929).[91] Within this structure, Gribbon's portrayal makes him unequivocally a figure of fun, and the filmmakers seem to have lavished particular attention on making him as absurd as possible. The script for *The Big Palooka* notes how "he [Gribbon] is dressed in the loudest checked suit of exaggerated cut. The brightest tie, bull dog shoes and hair slicked down on his brow. Thelma can only stare, open-mouthed and speechless"—a laugh-getting appearance that the writers for *The Lunkhead* subsequently capitalized upon, describing Gribbon as sporting "his Sunday-best Homer outfit (as worn in 'The Big Palooka')."[92] His first appearance in *The New Halfback* (November 1929) similarly marks the character's ridiculousness. A small-town boob, Elmer Buckley (Gribbon) is introduced to class on his first day at college, only to immediately start waxing nonsensical about his favorite pastoral fauna: "People can learn a lot from the whippoorwill," he instructs his peers. "He is a home-loving bird. I have often watched the papa whippoorwill. He starts out in the morning with a song on his lips"—whistles—"Mama Whippoorwill cheers him on"—whistles—"Where is he going? He's going out to get mama whippoorwill some nice big worms. In the summer time he brings smooth ones and in the winter he brings wooly ones."[93] For the remainder of the film his college teammates refer to him as "Mr. Whippoorwill."

It would be a mistake to assume that Gribbon's rube persona in any way contradicted the possibility of the films' heartland appeal. Rube stereotypes hardly spoke only to the prejudices of the big city: quaint, bib-overalled countrymen had been stock figures of the touring medicine shows that played in village squares and small-town opera houses in the South and Midwest from the late nineteenth century, and as film historian Charles Tepperman has shown, comic rubes also appeared in silent-era instructional films directly targeting a rural audience.[94] Perhaps this is why Gribbon's films seem to have won favor even among the communities he seemed to be ridiculing: "It's old stuff, but it made the farmers laugh on Saturday and if it's good enough for them it's fine for me," commented one Alabama exhibitor in 1935 after screening a reissue of *The Big Palooka*, the film that had done most to establish Gribbon's "palooka" persona.[95] Certainly, the sense in which rural populations may have been able to laugh at their own stereotypes was a frequent observation in studies of small-town culture from the early sound period. For instance, Albert Blumenthal's 1932 *Small-Town Stuff*—published under the auspices of the University of Chicago's famed Department of Sociology—commented explicitly on the willingness of small-town citizens to make fun of themselves. "The jests of city people at the expense of small towns are proverbial," Blumenthal wrote, "but what is not so well known is that alert small-towners are even more relentless in praising, condemning, and jesting about the small town"—a condition the author connected to a widespread fear of cultural backwardness and a desire to keep "up-to-date" with respect to "standards set by the larger cities."[96] Perhaps, then, for rural filmgoers, characterizations such as Gribbon's served an assimilative function by symbolizing gaucheries to be abandoned as they adapted to a modernizing nation. To laugh at their own stereotype would, from this perspective, have been a way for heartland audiences to negotiate the pressures of modernity against fears of backwardness.

Such a reading certainly makes sense of the role played by Andy Clyde's "Ed Martin" character, an eccentric but kindly patriarch whose role in these films is to mediate the opposed suitors. Typically, Mr. Martin begins by favoring Gribbon's suit for his daughter's hand, only to turn against the Gribbon character by film's end to endorse Thelma's choice. In *The Constabule*, for instance, he initially tells his daughter to forget about her college beau and marry Homer, the local constable ("That college put a lot of highfalutin' ideas in your head that you'll have to get out," he tells her in an early version of the script).[97] Those sentiments are reversed, however, when Homer mistakenly tries to arrest his future father-in-law on suspicion of theft, leading an infuriated Mr. Martin to chase him into the distance—and out of the film—with a rifle. In such instances, Mr. Martin actualizes a relation to modernity that may have relieved audiences of their anxiety at being outpaced in an urbanizing nation: his role in brokering Thelma's romance with her city-bred boyfriend secures a place of ongoing authority for small-town values, even as the films work to circumscribe and reject the rube-ish backwardness that becomes Gribbon's burden. What Mr. Martin represents might, in fact, usefully be seen as a kind of "provincial modernity," to borrow a phrase from film historian Kathy Fuller-Seeley. As developed by Fuller-Seeley, the notion of provincial modernity addresses the ways heartland America came to accept elements of modernity by "adapt[ing them] to provincial tastes" and values, thereby allowing "some modern ideas to slip quietly in" beneath the cover of traditionalism.[98] (Interestingly, her example of this process is cinematic narrative—specifically, the way early film genres like the western offered traditionalist period representations that nonetheless also gave scope to "modern" depictions of speed, consumerism, and gender equality.) But we might also think of the dynamics of provincial modernity from the other side; that is, not simply as a camouflaging of "modern ideas" but as a prouder affirmation of the local as the necessary filter and ultimate arbiter of modernity's effects, their promise and their problems. What one finds in the Clyde-Gribbon films is thus an acceptance of the new—as represented by Thelma and her city boyfriend—*only* through its concordance with the old, through the folksy and down-to-earth approval of an old-fashioned patriarch.

That Clyde did, in fact, appeal to provincial values was sometimes indicated in the exhibitors' comments from "What the Picture Did for Me." One Kentucky showman, for example, celebrated Clyde's comedies as "good old-fashioned slapstick" and noted elsewhere that "Clyde is a real comedian, even if he is not appreciated by the younger element."[99] Although Clyde's comedies were far from universally popular, they were evidently standouts for some small-town exhibitors who described them as "always pleasing" (Anamosa, Iowa) and "as good as any they make" (Dante, Virginia), using language that invoked their appreciation of Clyde's classic American rube-fool as wise man figure: "Andy Clyde always brings laughs to our rural lads," commented one.[100] The amazing longevity of his career further suggests how his comedy corresponded to the specific cultural field of sound-era slapstick. Although Sennett departed for Paramount in 1932, Educational retained Clyde's services for two seasons of *Andy Clyde Comedies,* before Clyde himself departed in 1934 for Columbia's short-subjects division, where he continued his "old man" characterization in seventy-nine starring shorts, until his departure from Columbia in 1956— the longest run for a single comic persona (as opposed to a comic team) in American film history.[101]

But Gribbon and Clyde were hardly the only Educational comedians to adjust their comic personas to the changing market for slapstick. Arguably the most remarkable of these adjustments was Buster Keaton's, in sixteen shorts released between 1934 and 1937 as part of Educational's new *Star Personality* series.[102] Keaton's path into the sound era had, of course, been famously troubled, culminating in his firing from MGM at the start of 1933, by which time his reputation as a hapless alcoholic had made him unemployable at any of the major studios. Yet Keaton's turbulent career trajectory had also witnessed some surprising shifts in his comic persona. Starting from his second feature at MGM, *Spite Marriage* (1929), Keaton began appearing under the character name of "Elmer" in his films, marking a turn away from the resourceful persona of his earlier features toward a more clueless, dim-witted characterization. Although the Elmer persona initially lacked a stable social identity—a dry cleaner in *Spite Marriage,* a rich milquetoast in *Doughboys* (1930), even a taxidermist in *What! No Beer?* (1933)—Keaton would streamline the characterization in the direction of small-town "boob" roles at Educational, in such rural-themed comedies as *One-Run Elmer* (February 1935), *Hayseed Romance* (March 1935), and *Grand Slam Opera* (February 1936), as well as the hillbilly farce *Love Nest on Wheels* (March 1937), among others. At least eleven of Keaton's Educational shorts feature some kind of country or small-town setting, and many of them push toward a style of comedy that integrates physical slapstick with character types derived from rural humor traditions (fig. 20).[103]

Nor can there be any doubt that Keaton's Educational films, like Clyde's, satisfied small-town exhibitors who called for a return to "good old-fashioned slapstick." "Good old Buster," wrote a Michigan exhibitor in the pages of *Motion Picture Herald.* "He's still the best pantomime comic on the screen." "Keaton's comedies are favorites [with my audience]," chimed in a theater owner from Clatskanie, Oregon, while a rural Kentucky exhibitor agreed, commenting that "Buster is always liked here."[104] *Palooka from Paducah* (January 1935), featuring an amusingly fake-bearded turn from Keaton as the youngest son in a hillbilly family looking to break into professional wrestling, received particular acclaim as a "comedy that is a comedy" (Eminence, Kentucky), "far above the average Educational" (Plano, Texas), and containing "more laughs than any comedy ever run" (Elvins, Missouri).[105] In fact, the only Educational product lines that approached the praise accorded Keaton's films by smaller exhibitors in the mid-1930s were, for the most part, even more explicit in their adaptation to rural settings and themes: for instance, the two series

FIGURE 20. In *Palooka from Paducah* (January 1935), Buster Keaton gathered members of his real-life family to play a clan of wrestling hillbillies. From left to right. his sister Louise, father Joe, mother Myra, and Buster himself. Courtesy Academy of Motion Picture Arts and Sciences.

featuring the blackface duo Moran and Mack, released between 1932 and 1934 (the second series abruptly terminated by Charles Mack's death), and most tellingly of all, the previously mentioned *Song and Comedy Hits* series of musical shorts. Equally popular, the eccentric dancing duo of Tom Patricola and Buster West were perhaps exceptions to this pattern, initially establishing their reputation in musical comedy shorts as a pair of love-happy, toe-tapping sailors. Yet their tenure as Educational headliners, in three series of six two-reelers between 1935 and 1938, nonetheless began to introduce telling variations: rural settings and character types began to appear with increasing frequency—as in *Happy Heels* (August 1936), the plot of which required them to "impersonate rubes" to invade a nightclub, or in *The Screen Test* (December 1936), which cast them as nimble-footed understudies in a "rural Little Theatre."[106]

*

We seem a long way from the urban, and frequently urbane, worlds of so many 1920s comedians, yet it would be a mistake to overstate the role of these developments in Educational's product. Certainly, Educational's output was never to any *thoroughgoing* extent reconstituted with a small-town public in mind. As with all shorts companies, variety remained the key, and throughout the 1930s, Educational sought a strongly diversified appeal for its product. The major trend in Educational's programming was in fact a continuing de-emphasis on live-action slapstick-style comedians (with only Buster Keaton really fitting the bill by the 1936–1937 season) and, correspondingly, a rising focus on musically oriented comedies (e.g., the success-ful West and Patricola shorts; the *Star Personality* releases featuring song and dance men like Pat Rooney, Herman Timberg, and a young Danny Kaye; and Jefferson Machamer's *Gags and Gals* pictures, part of Educational's two-reel *Musical Comedy* line). Indeed, for the 1937–1938 season, the use of separate series names for its two-reel comedies was discontinued altogether and the only official series to remain were the one-reel *Treasure Chests, Song and Comedy Hits,* and *Terry-Toons* cartoons.

Still, it would be no less of a mistake to dismiss Educational's rural-themed slapstick as, in consequence, a mere footnote to the history of short-subject com-edy, for to do so is to overlook the strategies the company adopted in constructing and anticipating an audience during the sound era. Nor was Educational in any way alone in this respect. Other companies that remained in the knockabout game seem to have similarly taken steps to reconstitute their comic output in part for traditionally "down-market" hinterland audiences. Consider, in this respect, the sound-era career of Charley Chase, once the most dapper of men-about-town comedians, who began appearing in rustic settings in the 1930s in a number of comedies placing him in hillbilly land. Chase's experiments with these formulas began at Hal Roach, with *The Real McCoy* (February 1930), *One of the Smiths* (May 1931), and *Southern Exposure* (April 1935), and continued into his association with Columbia—where he began work in 1937 as both performer and writer/director—with *Teacher's Pest* (November 1939) (fig. 21). The first of these, *The Real McCoy,* represented an obvious attempt to capitalize on the hillbilly music craze of the time by featuring a plot requiring Charley to prove his south-ern heritage via his musical abilities. Unusually for a short, it received advance notice from the *Herald*'s regional columnist J. C. Jenkins, who witnessed the film's production while visiting the Hal Roach Studios ("This one takes everything in the bake shop," Jenkins promised rural exhibitors).[107] While this style of comedy may have been somewhat out of the norm for the Roach studios, Chase's experi-ments with hillbilly humor were a much better fit at Columbia: as one of the little three, Columbia—like Educational—had no stable or guaranteed access to major-owned first-run houses and, in consequence, had long targeted the bulk of its product to hinterland tastes, not only in "B"-grade westerns starring Buck Jones and Gene Autry during the 1930s but also in rural-themed slapstick shorts. In addition to its long-running Andy Clyde series, Columbia's short-subjects department had a habit of using "fish out of water" plots that placed comedians in the midst of some hillbilly feud—for example, Swedish-dialect comedian El Brendel's *Ay Tank Ay Go* (December 1936) and *Love at First Fright* (July 1941),

FIGURE 21. City slicker Charley Chase awkwardly adapts to hillbilly life in *One of the Smiths* (May 1931). The coonskin cap is a skunk. Courtesy Academy of Motion Picture Arts and Sciences.

as well as Chase's aforementioned *Teacher's Pest,* which featured him as a city schoolmaster sent to teach the mountain folk readin', 'ritin', and 'rithmetic. Predictably, all of these comedies relied on exaggerated, rube-ish characterizations; still, the trajectory of the humor inevitably cut both ways, in gags that played equally on the tenderfoot clown's effete mannerisms as on stereotypes of hillbilly roughness. More than simply ambiguous, such comedies might more usefully be seen as bridging competing perspectives on the hinterlands, establishing basic comedic resources through which rural audiences may, as with the earlier Clyde-Gribbon films, have imaginatively negotiated the divided field of Depression-era mass culture.

It would be possible to continue listing examples, but the point should be sufficiently clear.[108] The evolutions of comic settings, formulas, and typology I have been tracing were not abstract, but took place within a divided exhibition market that prompted producers and distributors of short-subject slapstick—particularly independents, like Educational—to take full account of heartland audiences and their values. It is from this vantage point, in fact, that we can return one last time to the interpretations of slapstick's sound-era "decline" offered by critics like Agee and Kerr. For whereas they viewed slapstick's fate in largely aesthetic terms, this chapter has traced a complex interlocking of economic, industrial, and social factors to show that something more than aesthetics was at stake: that, ultimately, slapstick's critical fall from grace was tethered to the fading cultural capital of the heartland populations to which it increasingly spoke. Indeed, the very notion of aesthetic decline is less useful in this respect than what sociologist Pierre Bourdieu describes as "banalization," that is, the way cultural forms are devalued over time as their relation to their public changes. As Bourdieu notes, any innovation within a cultural field—and sound's impact on film comedy certainly counts as such—has the effect of attracting those audiences most concerned with distinction and cultural capital, while the once-popular forms (say, slapstick) consequently lose distinctiveness and witness their clientele age and the social quality of their public decline: "Thus the social ageing of a work of art, the imperceptible transformation pushing it towards the *déclassé* or the classic, is the result of a meeting between an internal movement, linked to struggles within the field provoking the production of different works, and an external movement, linked to social change in the audience."[109]

Banalization, in this sense, refers not simply to the process whereby a cultural trend or practice becomes outmoded, but to the material transformations that underscore or facilitate that process. Bourdieu's own (misogynistically framed) example of this process is perfume—specifically, how the great brand names forsake distinctiveness by mass marketing their product, thus driving away many of their original customers and leaving only a "composite clientele made up of elegant but ageing women who remain faithful to the perfumes of their yesteryears and of young but less wealthy women who discover these outmoded products when they are out of fashion"[110]—but it is clear that the field of short-format film comedy also fits the template. As we have seen in previous chapters, the innovation of sound was a catalyst for fresh hierarchies within the field of film comedy, pitting the sophisticated cultural capital of the new Broadway-style comic shorts against the more established slapstick style. What now becomes clear, however, is the way slapstick was in turn reconstituted as hokum, finding a new "composite clientele"—as well as new subject matter—among the middling sensibilities of small-town and heartland publics.

This, then, is the final context within which Educational's sound-era fortunes should be situated, and it suggests, by way of a closing observation, a further nuancing of slapstick's relation to that much-contested category of cultural experience, *modernity*. We have already seen in a previous chapter how slapstick's claims to a kind of vernacular modernism were, by the mid-1920s, significantly qualified by new patterns of metropolitan sensibility that disparaged the form as old hat. What now deserves to be stressed is how developments in the short-subject marketplace exacerbated this displacement by aligning slapstick with alternative vectors of cultural experience that took their cue from the conservatism of the heartland.

Rather, then, than adhere to the scholarly consensus that has hypostasized slapstick as a kind of aesthetic reflex of urban modernity, it would be more fruitful to insist instead on its variable social character as a form that addressed diverse popular logics. One possible logic, to be sure, related to city-bred experiences of class and ethnic division in the early twentieth century, as well as to the impact of changing technological regimes and mechanization—all of which indeed became tropes that defined slapstick's celebrated "modernity" for much of the silent era. Yet, by the 1930s, this chapter has argued, slapstick's cultural appeal would settle along a quite different divide: the growing distinction pitting metropolitan cultural hegemony against an assertive regionalism. And it is within the context of this profound relocation of slapstick's cultural place that the genre's decline—and Educational's history—finds an ironic horizon of interpretation: no longer urban but small town, not simply an "anarchic supplement" to technological modernity but a horizon for the more equivocal registering of provincial modernity.

But the final irony, for this chapter, is this: in 1934, an Educational release received the Academy Award for best novelty short, a surprising achievement perhaps, given that the company's output had been so thoroughly consigned to the margins of the distribution hierarchy. But the film was not a comedy; titled *Krakatoa* (April 1933), it was a three-reel science documentary showing the eruption of the undersea volcano. Finally fulfilling Earle Hammons's long-abandoned mandate, Educational's one outstanding critical success of the 1930s, the film that most notably received the stamp of a critical legitimacy otherwise withheld—that film really was an educational picture.

4

"I Want Music Everywhere"

Music, Operetta, and Cultural Hierarchy at the Hal Roach Studios

In their 1928 stockholders' report, the Hal Roach Studios' board of directors anticipated the company's conversion to sound with calm assuredness:

> The last few months has [*sic*] witnessed the advent of another element in the production field; that is, the talking or sound pictures. It is, of course, difficult to foretell what the eventual outcome of talking pictures will be or the eventual form they will assume. One thing is certain, however, that is that they are at the present time an element in the amusement field apparently having a definite appeal to the public, and properly handled, it promises to be a great addition to the entertainment value of pictures and a great aid to the producer in building up the interest in the picture intended. The company has placed itself in a position to gain by any and all new methods and devices introduced in the field.[1]

Confidence is to be expected in a stockholders' statement, but for the Roach Studios such an attitude likely came easy, at least in comparison with the competition. Unlike the independent producer-distributor Educational—which had initially flubbed its transition by backing the Vocafilm technology—the Roach Studios enjoyed the luxury of a financing and distribution deal with industry powerhouse Loew's-MGM and opted to follow the parent company's lead in entering the uncertain waters ahead. The Roach organization was, for example, included in Loew's-MGM's initial contract with Electrical Research Products, Inc. (ERPI), for installation of sound technology, signed May 11, 1928; and, like Loew's-MGM, Roach contracted with the Victor Talking Machine Company, a phonograph manufacturer and ERPI licensee, for recording equipment supply and the manufacture of soundtrack discs.[2] The arrangement with Victor brought not only equipment and technology but also new personnel. Victor employee Elmer Raguse arrived in late 1928 to prepare the soundproofing of the studio's existing stages, then signed on to serve as Roach's permanent sound engineer once the installation was complete. The following year, Leroy Shield, Victor's A&R man for the Western United States, was sent to the Roach lot in Culver City, where he remained to write musical cues and themes for the studio's comedies.[3] Roach's firm quickly availed itself of these new corporate and personnel resources to experiment with new forms for sound comedy that included music as a principle of comic form.

One of the distinctive aspects of the Roach Studios' passage through the conversion era, this chapter argues, was the role that music came to play in negotiating that transition. If sound's first and most direct contribution had been to open new possibilities for comic sophistication through *speech* (discussed in chapter 2), and if this in turn had been answered by knockabout's resurgence under the aegis of *noise* (chapter 3), then the use of *music* supplied Roach's organization with a third path that promised to salvage its product from the cultural and industrial marginalization against which all short-comedy companies struggled during this period. In focusing on these musical endeavors, moreover, the chapter aligns itself with emergent scholarly trends focused on the consolidation of film and music industries wrought by the transition to sound. Rather than approach that transition from a perspective focused narrowly on the *film* industry, the recent scholarly tendency has been to examine larger issues of sound technology *across media* during the 1920s and 1930s. Hollywood's conversion thus emerges not as an isolated technological shift but as the catalyst for a developing pattern of media industry convergence that saw the integration of radio, music publishing, and film businesses during these years.[4] As media historian Ross Melnick explains,

> In the span of five hectic years [from 1926 to 1930], family-owned film companies morphed into global entertainment giants that were both horizontally and vertically integrated with numerous media distribution channels and profit centers. By 1930, RCA, Paramount, Warner Bros., Fox, and Loew's were all convergent media conglomerates, with many owning motion picture production, distribution, and exhibition, music publishing and recording divisions, and broadcast networks, stations, and/or radio programs. There no longer was a walled-off "film industry" but rather an "entertainment industry" that produced motion pictures among a host of other products and media.[5]

The conversion to sound brought film and music-related industries together, creating fresh opportunities for studios to establish multiple media connections for incorporating and promoting musical content across several platforms. Warner Bros., for instance, acquired several large music-publishing companies (M. Witmark and Son, Harms Music Publishing, and others) and even developed a short-lived chain of music stores. Paramount began radio broadcasts from its Hollywood studio in 1928, and in 1929 temporarily acquired half interest in the CBS radio network. In the process, the "music sector" of the economy was profoundly transformed: music-related industries came to be concentrated in media centers like New York and Los Angeles, spurring a significant migration of musical talent, including people like Raguse and Shield. For example, between 1921 and 1929, the Los Angeles Local 47 of the American Federation of Musicians saw its numbers quadruple to about four thousand, as each of the major studios developed permanent music departments with contracted composers, arrangers, orchestrators, librarians, and resident orchestras, all working under in-house music directors.[6]

Given the marginalization of short subjects in most film historical research, one is not surprised to find that short-subject producers have been ignored in these developments, too. Yet a miniaturized version of these same convergent processes can be observed at the Hal Roach Studios, where they inspired an evolving range of approaches that incorporated music as an element of slapstick form. The company entered the talkie era in early 1929 with four short-film series—Laurel and Hardy, Charley Chase, Our Gang, and the Roach All-Stars— into which filmmakers soon began including Tin Pan Alley–style ditties as a potential basis for cross-media tie-ins (phonograph and sheet-music sales). The following year the studio institutionalized "wall-to-wall" (that is, continuous) background music across all its series—over a year before similar scoring practices gained traction in features—with scores composed by Leroy Shield and arranged for small jazz orchestra. In these ways, Roach's filmmakers translated an inherited culture of commercial popular music into innovative practices of background musical accompaniment. Yet that inheritance was in turn abruptly transformed when, beginning in the early 1930s, Roach decided to move into feature film production, abandoning the Tin Pan Alley idiom of the studio's earlier shorts to launch Laurel and Hardy in a series of feature-length Viennese-style operettas (e.g., *The Devil's Brother*, 1933; *Babes in Toyland*, 1934; and *The Bohemian Girl*, 1936).

This chapter explores the factors that gave rise to these different idioms of what can be called "slapstick musicality." Rather than simply catalog them, however, my intent is to indicate how changes in these idioms partook in slapstick's devaluation during this period. In exploiting a Tin Pan Alley idiom, Roach's filmmakers were working with the framework of a "distinctly modern art," in Ann Douglas's terms, whose commercial appropriation of vernacular traditions (African American ragtime, Irish ballads, Stephen Foster melodies) resonated with the broader landscape of metropolitan culture and Jazz Age nightlife.[7] In subsequently embracing European operetta as a format for the Laurel and Hardy features, Roach's filmmakers were opting for the safety of middlebrow strategies of appeal to offset the greater financial risks of feature filmmaking. The result, however, was to drastically resignify Roach's most popular slapstick team, which was now excised from the contemporaneity of vernacular idioms—both comedic and musical—and instead consigned to a mythic past: nostalgic tales of dashing brigands, gypsy revenge, and the restoration of aristocratic order placed Laurel and Hardy within what Susan Stewart theorizes as the "infinite time" of fairy tale.[8] What emerges from musical processes at Roach is, in this sense, a further mode of the "aging" of slapstick, one linked not only to changes in the form's audiences and the temporality of their tastes (as in banalization) but also to the defensiveness of production practices seeking the cachet of cultural forms that were already consecrated, already—like operetta—"classic."

"WHERE THEY THOUGHT THE MUSIC OUGHT TO BE": SONGS AND UNDERSCORING IN THE HAL ROACH STUDIOS' EARLY SOUND OUTPUT

Needless to say, none of these musical approaches emerged immediately; rather, they were arrived at only after a "feeling out" phase that encountered occasional dead ends. The initial approach to sound considered at Roach sought simply to maintain the status quo of the studio's established mode of production. At the end of November 1928—simultaneous with Elmer Raguse's arrival to prepare the sound installation—studio manager Warren Doane recommended that the studio's sound productions reserve dialogue only for later dubbing or close-up inserts. A testimony to the confusion and uncertainties provoked by sound, Doane's memo deserves quoting at length:

> It would be my idea that we should continue the making of silent pictures exactly as we have in the past up to the time when the picture has been previewed and finally accepted as ready for shipment. At that time I believe it will be possible in a very short space of time—not more than an hour or two—to photograph synchronized dialogue action, which then cut into one of the negatives will give us a dialogued motion picture . . . .
>
> The installation necessary to do this, in my opinion, would consist of a sound-proofed room in the stage, a sound-proofed projecting room also in the stage, and an appropriate movable monitory room, a sound track recording camera, together with the necessary microphone and mixing panel equipment, and a means of sound-proofing cameras.
>
> The dialogue which will accompany closeup [sic] action, in my opinion, should be made in the sound-proof room. The dialogue which will accompany the long-shot action should be made in the sound-proof projecting room and be recorded on film by the Movietone camera. In the case of closeup [sic] the characters being re-photographed at the same time recording is made; and in the case of longshots the sound recording added to the film already made.[9]

Doane's motivation was frankly economic (he wanted to keep the "cost of the studio installation ... at a minimum" and to avoid "loading the picture costs unnecessarily"), but his minimalist approach to the spoken word harmonized well with the comic philosophies of the slapstick filmmakers discussed in the previous chapter: dialogue was to be avoided as far as possible.[10] A year later, in fact, Roach himself advocated a similar position, touting the benefits of pantomime and sound effects over dialogue in the pages of *Motion Picture News:* "The art of pantomime is as old as amusement itself and there isn't the slightest chance that dialogue ever will entirely displace pantomime on the screen. Dialogue can't possibly take the place of pantomime in causing laughs .... [By contrast,] sound effects in pictures are going to find a definite niche in the market. There is no doubt about that."[11] Interestingly, something close to this was codified at the *Our Gang* unit, headed by director Robert F. McGowan, where the child actors' stilted delivery of lines in their first sound production, *Small Talk* (May 1929), led to drastic dialogue pruning for their second, *Railroadin'* (June 1929). As Robert Lynch, manager of Loew's-MGM's Philadelphia exchange, commented in a memo to Roach on this issue: "I think if the Gang comedies were just about 25% dialogue and 75% silent they would be a whole lot better, for it was the fast action that these kids could put over in a silent comedy that got the laughs. As matters now stand, it takes these kids too long to get their dialogue over to make the thing rapid fire enough, so why not try to make the next one along these lines?"[12]

There was little here to differentiate Roach's filmmakers from others wrestling with the challenge of sound, but more distinctive options were also being explored. One of these, touched on earlier, was to include songs as a basis for synergistic tie-ins—the idea being that Victor would release the performances as commercial records. Songs were, in fact, a feature of Roach's sound shorts from the outset: the very first sound release in the *All-Star* series, *Hurdy Gurdy* (May 1929), included scenes of Thelma Todd and Eddie Dunn harmonizing on "She Lives Down in Our Alley" and "My Gal Sal." The success of that pairing convinced Victor and Roach of commercial possibilities, setting in motion plans for a phonograph of a second Todd-Dunn duet, "Honey," for tie-in with *Dad's Day* (July 1929), and even a series of all-musical films, although both proved nonstarters.[13] Slightly more successful in synergistic terms were the early sound releases in the Charley Chase series, which commonly featured jazzy comic ditties, some penned by Chase himself. Early in 1930, one of these tunes—Alice Keating Howlett and Will Livernash's "Smile When the Raindrops Fall" from the comedy *Whispering Whoopee* (March 1930)—was released in sheet music form, seemingly prompting encouragement for further such endeavors from Loew's-MGM offices (fig. 22). Responding to a preview screening of the Charley Chase musical three-reeler, *High C's* (December 1930), exchange manager Lynch again took it upon himself to give feedback, writing to Roach that "the way that music worked in there is the last word. It was the

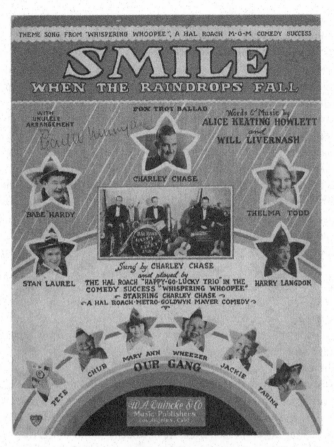

FIGURE 22. Sheet music for "Smile When the Raindrops Fall,"
featured in the Charley Chase comedy *Whispering Whoopee* (March
1930). Courtesy Vince Giordano.

opinion of about twenty others who saw this comedy that two or three more like
that would be just what the public wants (I mean that music worked into it)."[14]
After Roach replied by agreeing "to mix one of these musical type of pictures into
the Chase Series every once in a while," Lynch subsequently reiterated his point
in connection with the debut entry of the Thelma Todd–Zasu Pitt series, *Let's Do
Things* (June 1931): "You boys overlooked, in my opinion, a wonderful bet in not
having those singing voices in at least three times as much of that comedy as you
did . . . . I think the movie fans would be glad to listen to a whole half a reel of such
voices . . . particularly with good songs."[15]

Such an emphasis on song was broadly characteristic of the early sound period,
when the possibilities for music cross-promotion encouraged the Hollywood
studios to include songs in an unexpected variety of genres, including even

westerns and melodramas (as Donald Crafton has noted, "Around 1929–1930, it was the rare movie that was *not* a musical in some sense of the term").[16] But it was far more distinctive for producers of short-subject slapstick who, with the exception of major-affiliated companies like Roach, generally lacked the necessary musical talent and corporate ties to sustain such tie-in marketing.[17] Where Roach's early sound shorts most decisively innovated, however, was in their swift adoption of musical scores and themes. Here, acknowledgment must be given to two individuals who had arrived at the Hal Roach Studios by following the intersecting pathways linking the era's film, radio, and recorded sound industries: Marvin Hatley and Leroy Shield.[18] An Oklahoma-born substitute pianist at Warner Bros.'s radio station, KFWB, which had begun broadcasting in March 1925, Hatley had been hired to a new radio job for the Roach Studios' in-house station, KFVD, shortly before the studio's conversion to sound. Appointed musical director for the studio's sound releases, Hatley made an early mark by introducing one of the first examples of a musical signature in sound films, the famous "Dance of the Cuckoos" motif—also known as "Ku-Ku"—that played over the opening titles of Laurel and Hardy releases. As Stan Laurel later recalled, the tune had begun life as Hatley's hourly time signal for the Roach Studios' *Cuckoo Hour* radio show. "That originally was taken from a little radio station at the Roach studio . . . . We did it [first in *Brats* (March 1930)] for a laugh because it sounded cute. They liked it at a preview and we decided to leave it in."[19]

Soon, the studio's shorts were regularly incorporating full background music contributed by Minnesota-born pianist and composer Leroy Shield, who had first come to Roach as part of the deal with Victor. Shield was not himself responsible for the earliest background compositions in Roach's sound shorts: some of the studio's very first all-talking releases—including Charley Chase's sound debut, *The Big Squawk* (May 1929), and the second *All-Star* talkie, *Madame "Q"* (June 1929)—had incorporated wall-to-wall background compositions contributed and recorded by Gus Arnheim's legendary Cocoanut Grove band.[20] Studio management soon switched to an in-house model, however, and, by the late summer of 1930, Shield was supplying original tunes for use in Roach's shorts over their entire running time. The first film to feature Shield's compositions in this way was a Laurel and Hardy comedy, *Another Fine Mess* (November 1930), the extant continuity script for which stipulates that, for each of the three reels, "instrumental music is played offscene through[out]."[21] Nothing in extant production files, however, indicates the style of these melodies, which were scored not symphonically—as would eventually become the Hollywood norm—but rather after the fashion of orchestral jazz, consisting of around a dozen players: three strings (violins/cellos), three brass (trumpets/trombones), three winds (clarinets/saxophones), and a rhythm section (commonly comprising piano, banjo, drums, and double bass). (In a 1935 letter to Roach studio manager Henry Ginsberg, Shield requested orchestration consisting of violins, saxes, trumpets, a trombone, piano, guitar, bass, and percussion, including xylophone.)[22] Nor do the files indicate the range of Shield's jazzy compositions during his brief stint at Roach, which saw him compose some sixty-eight tunes—foxtrots, ballads, hurries, waltzes, the gamut—along with numerous brief cues during a burst of extraordinary creativity from late 1929 to his departure in 1931. As Shield scholar Piet Schreuders has noted,

> Shield probably set out to score specific scenes for specific purposes—for example, his beautiful love ballad "You Are the One I Love" was usually played behind love scenes, and something called "It Is To Laugh" behind comic scenes—but [sound editor Elmer Raguse] soon began to stuff each film sound-track with Shield's music, with little or no regard to its original purpose. Roach paid the composer a flat fee of about $200 per tune, and was able to use them as he pleased, whether Shield liked it or not.[23]

Raguse could do this, moreover, because, while at Roach, Shield never approached his compositions in a way that directly timed the progression of individual tunes to the unfolding of an action.[24] Rather, Shield's compositions were discrete melodies, each conceived and recorded individually, which were then arranged by Raguse in sequences over each two-reeler's duration, either using the entire tune or (in most instances) just segments. Effects cues would be used, but the music was generally linked to action primarily at a level of overall tone. The approach is evident from the cue sheet for Charley Chase's *Looser than Loose* (November 1930), the first reel of which includes a scene in which Charley endeavors to present an engagement ring to Thelma Todd in the face of a series of frustrations (fig. 23). What Raguse does for the sequence (comprising cues 10–13) is to take two of Shield's preexisting compositions and alternate between them: on the one hand, Shield's standard love theme "You Are the One I Love," a waltz consisting of a sixteen-measure introduction and a sixteen-measure melody that basically can be repeated as long as necessary; on the other, "Your Piktur" (referred to on the cue sheet as "Picktur"), which is a brief laugh cue, first introduced for the *Our Gang* short *Teacher's Pet* (October 1930) and introduced here whenever Charley's romantic entreaties are comically interrupted (figs. 24 and 25, auds. 1 and 2). In other words, "You Are the One I Love" establishes the base romantic mood, from which "Your Piktur" punctuates comic deviations. The first such occurs after Thelma, in close-up, opens the ring box to look inside: a point-of-view shot shows a frugal ring with tiny stone, while the soundtrack shifts to the braying two-bar laugh effect that begins "Your Piktur" (figs. 26 and 27). The same musical articulation recurs when Charley is interrupted by the phone: angrily, he picks up the receiver and yells "HELL-O!" only for a cut to reveal his boss on the other end of the line, again punctuated on the soundtrack by a shift to "Your Piktur," this time for four bars (figs. 28 and 29). Although neither of Shield's compositions was specifically written for this film, Raguse arranges them segmentally in a way that precisely articulates the momentum of the film's action (vid. 4). This approach can be linked to the silent era, when there existed a number of handbooks like Erno Rapée's 1924 *Motion Picture Moods,* consisting of melodies indexed under headings of mood or subject mat-

```
                    HAL ROACH STUDIOS, INC.
                    Culver City, California

                              Date:
                              Prod. No.  C-34

                   MUSICAL COMPOSITIONS RECORDED IN

       A Production entitled:  "LOOSER THAN LOOSE"

       Recorded by HAL ROACH STUDIOS INC. at Culver City, California

                         CODE FOR USAGE:

       BACKGROUND INSTRUMENTAL - BV        PARTIAL USAGE - P
       VISUAL INSTRUMENTAL     - VI        ENTIRE USAGE  - E
       BACKGROUND VOCAL        - BV
       VISUAL VOCAL            - VV
```

TITLE OF COMPOSITION	COMPOSER	PUBLISHER	HOW USED
Reel 1			
1. GANGWAY CHARLEY	LeRoy Shield	Unpublished	BI - P
2. GANGWAY CHARLEY	LeRoy Shield	Unpublished	BI - P
3. GANGWAY CHARLEY	LeRoy Shield	Unpublished	BI - P
4. BEAUTIFUL LADY	LeRoy Shield	Unpublished	BI - P
5. PICKTUR	LeRoy Shield	Unpublished	BI - P
6. BEAUTIFUL LADY	LeRoy Shield	Unpublished	BI - P
7. PICKTUR	LeRoy Shield	Unpublished	BI - P
8. GANGWAY CHARLEY	LeRoy Shield	Unpublished	BI - P
9. PICKTUR	LeRoy Shield	Unpublished	BI - P
10. YOU'RE THE ONE I LOVE	LeRoy Shield	Unpublished	BI - P
11. PICKTUR	LeRoy Shield	Unpublished	BI - P
12. YOU'RE THE ONE I LOVE	LeRoy Shield	Unpublished	BI - P
13. PICKTUR	LeRoy Shield	Unpublished	BI - P
14. RIDING ALONG	LeRoy Shield	Unpublished	BI - P
15. THE ONE I LOVE BEST	LeRoy Shield	Unpublished	BI - P
16. YOU ARE THE ONE I LOVE	Le Roy Shield	Unpublished	BI - P
17. CONFUSION	LeRoy Shield	Unpublished	BI - P
18. THE ONE I LOVE BEST	Le Roy Shield	Unpublished	BI - P
19. CONFUSION	Le Roy Shield	Unpublished	BI - P
20. TUNE	LeRoy Shield	Unpublished	BI - P
21. DANCING GIRL	LeRoy Shield	Unpublished	BI - E
22. TUNE	LeRoy Shield	Unpublished	BI - P
Reel 2			
23. TUNE	LeRoy Shield	Unpublished	BI - P
24. TUNE	LeRoy Shield	Unpublished	BI - P
25. GANGWAY CHARLEY	LeRoy Shield	Unpublished	BI - P
26. GANGWAY CHARLEY	LeRoy Shield	Unpublished	BI - P
27. IT IS TO LAUGH	LeRoy Shield	Unpublished	BI - E
28. ANTICIPATION	LeRoy Shield	Unpublished	BI - E
29. ANTICIPATION	LeRoy Shield	Unpublished	BI - P
30. ANTICIPATION	LeRoy Shield	Unpublished	BI - E
31. ANTICIPATION	LeRoy Shield	Unpublished	BI - P
32. LITTLE DANCING GIRL	LeRoy Shield	Unpublished	BI - P
33. YOU'RE THE ONE I LOVE	LeRoy Shield	Unpublished	BI - P
34. LET'S GO	LeRoy Shield	Unpublished	BI - P
35. ANTICIPATION	LeRoy Shield	Unpublished	BI - P
36. CONFUSION	LeRoy Shield	Unpublished	BI - E
37. GANGWAY CHARLEY	LeRoy Shield	Unpublished	BI - P

(Note: World Picture rights on all music used in scoring this picture is owned by Hal Roach Studios)

FIGURE 23. Cue sheet for the Charley Chase comedy *Looser than Loose* (November 1930). Courtesy Cinematic Arts Library, University of Southern California.

ter: accompanists would simply draw upon these to present an appropriate musical analogue for the developing action on screen. In effect, what Raguse does here is to sequence Shield's compositions in a fashion akin to how a silent film accompanist would have used Rapée's *Motion Picture Moods*, treating Shield's melodies as an ever-growing library of prerecorded cues that could be selected from and "plugged in" to appropriate scenes.

AUDIO 1. Melody from "You're the One I Love."

To listen to this audio, scan the QR code with your mobile device or visit
DOI: https://doi.org/10.1525/luminos.28.1

FIGURES 24–25. Transcriptions for "You're the One I Love" and "Your Piktur." Courtesy
Mauri Sumén.

AUDIO 2. Melody from "Your Piktur."

To listen to this audio, scan the QR code with your mobile device or visit
DOI: https://doi.org/10.1525/luminos.28.2

This treatment of his music disgruntled Shield, who around this time began complaining to his publisher that Roach's filmmakers were reusing his compositions in films for which they were not originally written and not paying him for the additional use. "I do not know just how to get around this," he wrote. "I cannot very well insist that music for each picture must be new, or that I must do the score, but I am wondering if in some way there might be a means of restricting their using any of the music without your permission as publisher, or mine as composer. In other words I do not like to lose this income."[25] By June 1931, Shield quit Roach to once again follow emerging corporate pathways, this time taking a job as musical director at the Chicago studios of NBC (also owned by Victor after merging with RCA). In Shield's absence, Roach's filmmakers continued to rely upon his preexisting compositions in new shorts, even using them for reissues of early comedies that

FIGURES 26–29. Musical sequencing in *Looser than Loose* punctuates the visual punchlines (cues #10-13). Thelma's engagement ring disappoints (26, 27) and Charley inadvertently yells at his boss (28, 29). The second shot in each pair is accompanied by Leroy Shield's laugh effect, "Your Piktur."

VIDEO 4. Clip from *Looser than Loose*.

To watch this video, scan the QR code with your mobile device or visit
DOI: https://doi.org/10.1525/luminos.28.6

had originally been released without underscoring (which is why the brief Shield tune "Bells," for instance, ended up being featured twenty-four times in Laurel and Hardy shorts alone). There was an attempt to bring Shield back to Roach in 1932 to contribute fresh cues (Henry Ginsberg wrote Shield in February to request "about twelve numbers consisting of six good Fox Trots, a couple of Hurrys, two Ballads, a Waltz and a Heavy Number"), although nothing came of it.[26] Shield did return briefly to help adapt Daniel Auber's score for the feature-length operetta *The Devil's Brother* in 1933, and was subsequently invited back by Ginsberg to head a new musical department at the studio in 1935.[27] Existing correspondence shows that Shield leapt at this new offer, but it appears that Ginsberg had overstepped his authority. Shield's replies to Ginsberg went strangely unanswered, and he instead received a terse and disingenuous rebuttal from Roach himself: "Our program for next year will be composed of feature length comedies," began the two-sentence letter Roach wrote to Shield early in 1936, "and as we have no musicals in this, will have no need for a Musical Director. However, if our policy ever changes, will be glad to have you with us."[28] Roach was not exactly unknown for such shabby treatment of his talent—he had summarily shown the door to comedian Snub Pollard in 1924 and would do the same to Charley Chase in 1936—and he likely realized that Shield was by this point somewhat superfluous to the studio's needs. Marvin Hatley had stepped into Shield's shoes to supply new cues as early as the 1933–1934 season, and at considerably less pay—two hundred dollars a week— than Shield would have required. Roach did soften to the point of signing Shield for a two-week period in mid-1936 to compose tunes for the Laurel and Hardy feature *Our Relations* (1936) at a salary of five hundred dollars per week.[29] But this was the last employment Shield ever saw at the Roach lot. Shield himself had become disposable even as his compositional model remained indispensable.

*

To see the stakes in Shield's musical practice more clearly, it is worth locating the Roach Studio's introduction of background music against broader scoring trends in Hollywood during the early sound period. In her book *Saying It with Songs,* Katherine Spring examines the classical Hollywood background score as something that began to take shape, through the early 1930s, out of the tail end of the theme-song craze of sound's earliest years.[30] By encouraging the use of songs willy-nilly in a wide variety of genres, Spring contends, the theme song trend had generated unmanageable problems for classical norms of narration and, by around 1930, theme songs rapidly began to fall from favor. This shift in musical practice was then accompanied by a change in the staffing of music departments as the studios bought off the contracts of the Tin Pan Alley songsmiths they had hired, instead retaining composers who brought a more European/classical style of orchestration to their scoring.[31] What's distinctive about Roach here is really three-fold. First, the institutionalization of underscoring occurred at Roach prior to *any* other producer of live-action fiction films. This is not to suggest that the Roach shorts were the first live-action all-talking films to include underscoring on this scale—Warner Bros.'s 1929 feature *The Squall,* released two weeks before Charley Chase's *The Big Squawk,* provides an earlier known case—but they were the first films for which the practice was standardized across a studio's entire product line.[32] Second, the approach to underscoring at Roach was conceived *within* a Tin Pan Alley mode, in contrast with the late Romantic idioms practiced and popularized by subsequent composers like Max Steiner, Erich Wolfgang Korngold, and Alfred Newman, with whose names the development of the classical Hollywood score is more typically discussed.[33] As such, the Roach films show how Hollywood underscoring practice initially accommodated a popular musical paradigm that bore scant resemblance to the symphonic—indeed, Wagnerian—model more commonly prioritized in scholarly studies of early sound-era compositional practice.[34] Third and finally, the need to maintain classical norms of narration can't actually explain the introduction of underscoring at Roach because slapstick, historically, has been a site of deviance vis-à-vis those norms.

The key factor behind the sustained adoption of underscoring at the Hal Roach Studios instead emerges from the specific difficulties that film comedians had faced following sound's advent, particularly as these concerned the issue of pace. As discussed in the previous chapter, the coming of sound had significantly impacted the pacing of slapstick action by making filmic time commensurate with profilmic reality: the standardization of motorized cameras and projectors ensured that films were now projected at the rate at which they had been shot.[35] The result, however, was to produce what one exhibitor deemed a "stultifying effect" on comedy by eliminating the undercranking effects on which the buoyancy of silent-era slapstick action had depended.[36] Comedy producers had, from the beginning, struggled to find ways around this difficulty: one technique— heavily employed, in fact, at Roach for the 1932–1933 *Taxi Boys* series—was to step-print sequences of comic action to create a "sped-up" feel. The chief avenue pursued at Roach, however, was to explore the possibilities of music as a means of *temporalizing* the image (the term is sound theorist Michel Chion's)—that is, to use underscoring as a means of dynamizing the visual action through musical qualities of rhythm and melody.[37] And here one finds that the studio's *comedians* took the lead in advocating this function for background music, as clearly indicated by the recollections of Marvin Hatley. "Every time Stan [Laurel] worked on a picture, he'd say, 'I want music everywhere.' He said, 'I do lots of pantomime, and I've got to have something going back there. A good, fast-paced music makes my stuff go better; if you take the music away, I don't move so fast.'"[38] Hatley also recalled that it was performer-directors like Laurel and Charley Chase who would give "the general idea of . . . where they thought the music ought to be," noting "Laurel usually wanted 100% music."[39] Unlike the initial use of songs at Roach, then, it was not the economics of convergence that motivated the dissemination of underscoring at Roach; rather, the direction of innovation was shaped by comedic norms of pacing and tempo inherited from silent-era filmmaking—as well as from the stage, where musical accompaniment for pantomime performances was standard—and that continued to govern the creative labor of performers like Laurel through the conversion period.

The point emerges clearly from a brief comparison of slapstick routines from the Laurel and Hardy shorts before and after the implementation of Shield's compositions. A good example of the former is the five-minute sequence from the team's second sound short, *Berth Marks* (June 1929), in which the boys undress and prepare for bed while crammed into an upper berth. Punctuated only by minimal (and apparently improvised) dialogue and accompanied by the sound of rail tracks, the routine provoked disappointment on the part of the *Motion Picture News* critic who described it as "monotonous": "The comedy consists for the most part of medium and closeup shots of the pair in the agonies of undressing. There's nary a variation."[40] It is thus significant that a parallel sequence from the Shield-scored *Be Big* (February 1931) did not receive such critique, since this suggests the difference that scoring could make. *Be Big*'s centerpiece is an even lengthier routine—a thirteen-minute set piece in which Babe, aided by Stanley, tries to remove a pair of boots in his hotel room— only this time with a Shield-composed score comprising fragments from the following melodies: "Ah! 'Tis Love," from the *Boy Friends'* two-reeler *Doctor's Orders* (September 1930); "Excitement" and "Hunting Song," both originally composed for *Pardon Us* (1931); "Rockin' Chair," whose appearance in *Be Big* marks its first; and "Slouching," one of the most widely used of Shield themes. Within this musical constellation, the plodding 3/4 tempos of "Rockin' Chair" and "Slouching" serve equivalent expressive functions, both used more or less interchangeably to churn up the tension of slow-burn physical routines (for instance, when Stan and Babe each get a foot caught in the other's sweater and are unable to extricate themselves); the lurching, triplet-based syncopations of "Excitement" are reserved for outbursts of frenetic energy (e.g., when Babe tears a curtain pole off the wall in frustration); and the upbeat, xylophone-driven "Ah! 'Tis Love" punctuates the two moments of respite when Babe thinks he's figured the boots out. (Based loosely on "A-Hunting We Will Go," Shield's "Hunting Song" seems primarily to have been chosen because of the film's plot: the boys are trying to trick their wives so they can attend a party at a hunting lodge.) "Temporalization" in this sequence is in fact less a mere matter of "fast-paced" music—as Laurel's own words would seem to imply—than, more ambitiously, of the use of unequally rhythmic melodies to structure the ebb and flow on which this chamber slapstick (two men, one pair of boots) depends.

Here, then, is a conception of scoring as a fragmented assemblage, not a Romantic unity, whose system is governed less by narrative values (only "Hunting Song" resonates narratively) than by the phenomenology of slapstick, by the rhythm and tempo of its physical action. As such, Shield's work at Roach confirms the suspicions of those cinema scholars and musicologists who have disputed the salience of Romanticism to the initial development of studio-era Hollywood scoring. As Jennifer Fleeger has argued, the conversion-era score was "disparate . . . rather than unified"; it was an "amalgamation of melodic fragments" whose conceptual basis was not in the unities of classical music, as most scholarship has assumed, but in the compressed, hook-heavy style of Tin Pan Alley.[41] What is perhaps even more of an affront to conventional histories is the fact that underscoring first became institutionalized in Hollywood not in dramatic filmmaking but in the specific field of *slapstick* cinema, for it was not in the name of classical narrative values that underscoring was first normalized but rather as a kind of affective "gel" that lent its qualities to the sensory registers of comedic pacing.[42]

"STRAINING THE RISIBILITIES": THE HAL ROACH
STUDIOS' TRANSITION TO FEATURES

We will approach subsequent developments in this paradigm of slapstick musicality by first turning to the economic and industrial factors that prompted the Roach Studios' transition to features. From any standpoint, the Roach lot was uniquely positioned to weather the storms within the short-subject market of the 1930s. Coddled by his distribution deal with Loew's-MGM, Roach had guaranteed access to a major-studio-owned theater chain that protected his product from the front lines of the battle against double features. Roach himself considered that deal "the smartest business move I ever made," and for good reason: he was the only independent producer of short comedies whose company survived the decade.[43] Still, the Roach lot could hardly stand entirely outside the fray. MGM may have been—famously—the most "Depression-proof" of the Hollywood majors; even so, the economic turbulence of the short-subject market ensured that Roach's profits occasionally transformed into losses. The Roach Studios had posted profits of $87,085 for its first full season following the Depression—July 27, 1930 to August 29, 1931—but then registered losses of $41,875 for the following six months.[44] The seesawing continued throughout the early 1930s: profits of $218,155 for the 1933–1934 season were answered by losses of almost the same amount, $221,919, the subsequent year, with more red ink coloring the ledgers for the 1935–1936 season.[45] Roach's first response to these market uncertainties had been to experiment with a three-reel format in a number of his series in the 1930–1931 and 1931–1932 seasons. The plan seems to have been to seize a "prestige" identity for the studio's releases that would differentiate the Roach brand from competitors in the short-subject field.[46] But the experiments failed to satisfy exhibitors—who complained that the films seemed "padded"—prompting a further change in strategy on Roach's part: seeing no future in shorts, he opted to transition gradually out of the short-subject market altogether to produce features. It was within those features that a new role for music was assayed.

Not that the Roach Studios' earliest dalliances with the feature format reflected any grand plan. The studio's first feature effort, *Pardon Us*, seems to have begun life as a regular Laurel and Hardy two-reeler that Roach's filmmakers, perhaps emboldened by their concurrent three-reel experiments, opportunistically extended to feature length. "As far as I can remember," Roach later recalled, "*Pardon Us* was supposed to be a two-reeler."[47] The idea for the short had been to use the massive sets recently built for MGM's *The Big House* (1930) as the basis for a prison-themed comedy. When the set-use arrangement fell through, however, Roach was forced to create new prison sets on his own lot (reportedly based on actual photographs of Sing Sing and San Quentin).[48] To offset the expense, Roach and his writers simply extended the film to six reels for release as a feature, rightly trusting that they would be able to convince MGM to release it. On May 25, 1931, MGM bucked its long-standing policy of only distributing in-house features and drew up a one-time deal to cover the release of *Pardon Us*.[49] The resulting feature did terrific business, grossing more than a half million dollars in the United States and Canada alone—a figure that Laurel and Hardy's subsequent features never managed to top.[50] Roach was now in a position to negotiate with MGM in advance for future features, although MGM permitted these arrangements only on a picture-by-picture basis and initially limited these deals to Laurel and Hardy vehicles; moreover, the comedy team was to continue their work in short subjects at the rate of around a half dozen per year. A deal for a second feature, *Pack Up Your Troubles* (1932), was signed on January 5, 1932, and for a third, *The Devil's Brother*, at the end of the year—at which point MGM doubled down on its commitment by contracting for two Laurel and Hardy features for each of the 1933–1934 and 1934–1935 seasons.[51] By mid-decade, the ongoing success of these films empowered Roach to begin to phase out his short-subject lines and place studio operations on a more consistently feature-length footing. Laurel and Hardy made their last short together in 1935, *Thicker than Water* (March), and then appeared exclusively in features for Hal Roach for the remainder of the decade, two per season; the Thelma Todd and Patsy Kelly shorts were abruptly terminated by Todd's death, also in 1935, at which point Roach leveraged Kelly into feature-length roles in *Kelly the Second* (1936) and *Pick A Star* (1937); Charley Chase was unceremoniously shown the door after his first solo starring feature, *Movie Night* (1936), misfired and had to be cut down for release as a two-reeler, retitled *Neighborhood House* (May 1936); finally, the *Our Gang* releases were phased down to single reels, "too successful to stop," in historian Richard Roberts's words, until Roach sold the whole franchise to MGM in 1938.[52]

To have convinced MGM to accept a regular supply of independently produced features was no small feat; as Richard Lewis Ward notes, an examination of MGM's release schedule in the 1930s shows that the studio accepted only three other third-party features during the entire decade, one of which was David O. Selznick's *Gone with the Wind* (1939).[53] But securing distribution was only part of the difficulty. The studio's filmmakers also had to negotiate the pitfalls of adapting the Roach brand of slapstick to the longer format.[54] It is moreover clear from contemporary reviews that Laurel and Hardy's initial features struggled precisely on this issue. *Pardon Us,* as noted, began life as a two-reeler that Roach's filmmakers then stretched out for six reels for release as a feature. Yet such an approach all but guaranteed a heavily episodic structure that diverged from classical norms of plot construction: as in many of their shorts, Laurel and Hardy are simply characters to whom things "happen"—here, imprisoned, caught up in an escape, recaptured, caught up in another escape—rather than goal-driven protagonists capable of sustaining an unfolding narrative.

Few reviewers harbored any illusions that the film was anything but a drawn-out short, and many reiterated the accusations of padding that had accompanied the studio's three-reel releases. "The material does not provide the consistent riotous fun that their short films do," observed one reviewer; "[The] full-length film is something of a strain on the partnership," commented another.[55] Similar complaints attended Laurel and Hardy's second feature-length effort, *Pack Up Your Troubles*, released the following year. Even though Roach's writers this time sought a stronger basis in story—the plot apes Chaplin's *The Kid* (1921) by casting Stan and Ollie as surrogate fathers to the orphaned daughter of their deceased war buddy—critics remained dubious about the duo's suitability for six-reelers. "This kind of entertainment, rough and comic, needs to be short, pungent and, above all, speedy to be really effective," Marguerite Tazelaar commented at the *New York Herald-Tribune*. "Two reels packed with it make for the kind of side-bursting hilarity Chaplin and Mack Sennett have been able to achieve. Six reels tend to strain the risibilities."[56] More striking than this, however, was the degree of hostility leveled against the film's slapstick sequences, which, some felt, were jarringly mismatched with the otherwise sentimental storyline. "They have allowed their low comedy to become entwined with the wrong kind of pathos," commented a reviewer for the London *Times*, criticizing in particular the slapstick depiction of trench warfare early in the film, when the young girl's father loses his life. "The soldier whose last hours are made miserable by a genuine concern for his motherless child ought never to have been enlisted in an army that already had Laurel and Hardy in its ranks."[57] Also jarring was an elaborate pie fight sequence—eventually cut from the finished film—which ensues when the boys, back from the front, disrupt a wedding. As scripted, the sequence seems to have been intended as an homage of sorts to one of the most celebrated sequences of the pair's silent shorts, the hugely elaborate pie fight from *Battle of the Century* (December 1927), which had involved the entire day's output of the Los Angeles Pie Company. (The screenplay for *Pack Up Your Troubles* makes the link to the earlier film explicit: "This develops into a 'Battle of the Century,' ending with the bride crowning the groom with the big cake.")[58] But lightning failed to strike twice, and a poor preview screening for *Pack Up Your Troubles* resulted in the sequence being pulled. As Felix Feist of MGM wired Roach executive Henry Ginsberg following the August 1932 preview, "Practically every preview card expressed repulsion at pie throwing episode which Hal agreed should come out."[59] The critical double bind was thus apparent: where *Pardon Us* received complaints for its *lack* of an integrative narrative, *Pack Up Your Troubles* provoked "repulsion" for slapstick sequences that *mismatched* its narrative frame.

Curiously enough, Laurel and Hardy had by this time already cameoed in another, quite different type of feature that may have suggested to Roach a more promising path for multiple-reel success. The film in question was the 1930 MGM operetta *The Rogue Song*, a high-profile, two-color Technicolor showcase for Metropolitan Opera star Lawrence Tibbett, directed by Lionel Barrymore, for which Laurel and Hardy had been loaned from Roach to appear briefly as "burlesque desperadoes."[60] Loosely adapted from Franz Lehár's 1912 operetta *Gypsy Love*, *The Rogue Song* had been a flagship for sound cinema's promise as a medium of musical uplift. Tibbett thus spoke of the film as an attempt to disseminate high culture for the American public: "Just as the radio educated the public in the matter of good music," he noted, "so the talking picture will show the beauties of music combined with dramatic action."[61] The movie was received in kind by a number of critics who considered it an "epoch-maker" that promised to fundamentally transform sound cinema's position within the field of cultural production: *The Rogue Song* "will surely start a concerted raid on the grandest opera houses in the world to get singers from the screen," one critic wrote; "the picture should open the door to great voices and really great composers," commented another.[62] But what were Laurel and Hardy doing here? Although their inclusion in the film may seem incongruous, most stage operettas did in fact contain comic foils to the romantic couple. As Donald Crafton has suggested, moreover, Laurel and Hardy's presence in *The Rogue Song* seems to have been calculated to ensure a broad, mass public for a feature that was otherwise marketed in distinctly highbrow terms.[63] Within the trajectory of Hal Roach's developing production strategies, then, it is possible that *The Rogue Song* offered an intriguing object lesson in how to negotiate the terrain of feature-length filmmaking. Casting Laurel and Hardy as comic relief in operetta-style plots would not only bring the pair's features in line with the prestige brand identity of Roach's distributor, Loew's-MGM; it also suggested a strategy for resolving the double bind that the team had encountered in their first two starring features. Such an approach would relieve the comedians of the burden of carrying a feature-length narrative—since they could now appear as auxiliary comic characters within otherwise serious romantic plots—and it would also provide a genteel framework for couching the duo's "low" slapstick. At a time of slapstick's waning mass appeal, the operetta model would enable Roach to identify the Laurel and Hardy features with a *diversity* of attractions: not just slapstick but romance and period settings, not just "lowbrow" comedy but "highbrow" opera. Accordingly, for their third feature, Roach aped the *Rogue Song* template by casting the duo in *The Devil's Brother*, based on Daniel Auber's 1830 opéra comique *Fra Diavolo*, and would do so again in two of their subsequent features: *Babes in Toyland* (based on Victor Herbert's 1903 operetta) and *The Bohemian Girl* (based on Michael Balfe's 1843 opera). In the process, Roach's filmmakers would not only completely invert the Tin Pan Alley paradigm of slapstick musicality first developed in shorts; they would also consign the studio's leading clowns to a realm of fairy-tale nostalgia.

"A CHILD'S CONCEPTION OF FAIRY LAND": SOCIAL ORDER AND FAIRY-TALE UTOPIA IN THE LAUREL AND HARDY OPERETTAS

To understand this process it is useful to attend first to what philosopher Jacques Attali has termed the "political economy" of music. Such an economy, Attali argues, revolves in the first instance around distinctions between what is considered *music* and what is *noise* in any given period; that is, how and where those distinctions are located and negotiated within social experience and cultural expression. Those distinctions, moreover, pertain to questions of social and political ordering. Noise is the very "image of subversion," Attali argues, to which music responds as a force of order, both a symbol and a tool of social cohesion.[64] "All music, any organization of sounds is then a tool for the creation or consolidation of a community, of a totality . . . . *Every code of music is rooted in the ideologies and technologies of its age, and at the same time produces them.*"[65] What Attali proposes, in short, is nothing less than a historical hermeneutic that comprehends music—traditionally, for aesthetic theory, the most ineffable and immaterial of cultural forms—as in fact in reciprocal interaction with material social formations. Musical forms thus not only correspond to or "reflect" concrete social forms, Attali suggests, but also play a productive role *within* the social process, whether as a means of actively reinforcing social or cultural order or as a provocation toward material social changes to come. (Tonal music, for instance, grounded as it is in a scientific ideology of harmony, thus becomes for Attali a bourgeois substitute for religion in imaginatively convincing people of the ideal of a harmonious and rationalized social order from which dissonance will be removed; free jazz, meanwhile, is viewed as the "refuge of a [revolutionary] violence" that as yet lacked a "political outlet.")[66]

This sense of music's role in reproducing social forms is important to us in this analysis, since it suggests a way of framing the Roach Studio's toggling between musical styles (Shield-style pop jazz, European operetta) in terms of the different social imaginaries they sustained. In accommodating popular musical paradigms as a basis for underscoring practice, Roach's output was initially configured in relation to musical idioms that corresponded, within the era's cultural imaginary, to a heteroglot conception of society. To take jazz as an example: a common image for the form was given in Paul Whiteman's musical film for Universal, *King of Jazz* (1930), which posited the music's origins in the combined influences of all manner of immigrant communities (albeit notoriously, in this film, with the omission of African Americans). What jazz meant in this context was notoriously slippery (music historian Krin Gabbard notes that "jazz" in the 1920s meant basically any fast-paced popular tune); still, the sense that it amounted to a kind of "industrialized folk-music," as the *Nation* dubbed it, was commonplace.[67] The same held for a range of cognate musical idioms, from ragtime to Tin Pan Alley commercial pop, which exhibited a "multicolored musical and linguistic palette" that appropriated the syncopated complexities of African musical tradition to Euro- and Anglo-American song structures like marches and ballads.[68] "Most of the star Tin Pan Alley composers and lyricists were Jewish and of recent immigrant stock," Ann Douglas writes, "and they found their finest interpreters in Negro as well as Jewish musicians and performers—Sophie Tucker, Fanny Brice, Al Jolson, Louis Armstrong, and Ethel Waters."[69] In the opinion of a host of commentators, the vernacular mixing that produced these various musical forms thus gave the nation its only fully developed "folk" art. (Or, as George and Ira Gershwin put it in the title of one of their early tunes, "The Real American Folk Song (Is a Rag).") It is thus notable how this hybridized musical imaginary seems to have informed the social content of the Roach Studios' early sound output: the popular tunes and jazzy ditties that commonly appear in the narratives of these early shorts often become fulcrums around which the studio's clowns forge communal connections across distinctions of ethnicity, region, and even nation in a kind of comically enacted utopianism. A remarkable instance is provided by the very first sound release in Roach's *All-Star* series, *Hurdy Gurdy,* in which sound is exploited both for ethnic dialect humor and for singing. Directed by Roach himself, the film is less narrative than slice-of-life, cutting between various scenes on the fire escapes of a New York City tenement as the residents sweat out a heat wave. Dialogue here works to establish a comedic tapestry of interethnic miscommunication and frustration: Edgar Kennedy, performing his familiar policeman role in a thick Irish brogue, vainly tries to get some sleep while harassed by his Italian neighbors' pet monkey; two German-Jewish neighbors, played by Max Davidson and Oscar Apfel, vainly try to discuss politics in broken English; all, moreover, become suspicious of their beautiful neighbor (Thelma Todd), who continually calls on the ice man (Eddie Dunn) to deliver blocks of ice. Popular song, by contrast, expresses a more integrative ideal; in counterpoint to the vignettes of squabbling neighbors is a series of scenes between Todd and Dunn, whose developing romance is conveyed through duets of "She Lives Down in Our Alley" and "My Gal Sal." Their performance of the latter provides resolution and the film's conclusion: Todd allays her neighbors' suspicions by explaining that she is a vaudeville animal trainer who needs the ice for the seal she has hidden in her bathroom; Kennedy apologizes "on behalf of mesself and all those furrners out there in the alley"; and Todd and Dunn harmonize one final time on the fire escape.

This model of a mixed musical community would return throughout the early period, notably in some of the Charley Chase musical shorts—for example, in the wartime comedy *High C's,* in which Chase fakes an armistice so that he can reach across enemy lines and enlist a talented German tenor into his close harmony quartet, or in one of his rural slapsticks, *The Real McCoy* (February 1930), in which Chase's city slicker proves his worth to distrustful hillbillies by leading them in an up-tempo rendition of the traditional song "Naomi Wise." It would also underpin the short-lived series of "comedies with music" launched in the 1933–1934 season, featuring Billy Gilbert and Billy Bletcher as the German "Schmaltz brothers." In this series, popular music greases the wheels of business success and assimilation for the ethnic clown. The first in the series, *Rhapsody in Brew* (September 1933), has the brothers tricked into buying a failing beer garden that they nonetheless turn into a hit nightclub destination by hiring an all-female jazz orchestra. In the third, *Music in Your Hair* (June 1934), the comedians play not brothers but German neighbors who put their battles aside when their Americanized children—the Bletcher character's daughter and Gilbert's son—fall in love and form a nightclub musical duo. Making visible the historical role of popular music as a vehicle for ethnic participation in America's mass culture, these comedies translate that process into utopian depictions of bottom-up communal formation and kinship.[70]

But this polyglot utopianism vanished with the Roach Studios' transition to operetta, which replaced the vernacular ideal of a mixed community with the depiction of a top-down political order. In his book *The American Film Musical*, Rick Altman discusses the operetta (which he dubs the "fairy-tale" musical) as an allegory of governance. Characteristic of the operetta—particularly in the Viennese tradition pioneered in the 1870s and 1880s by Johan Strauss II and continued into the new century by Franz Lehár—is the "equation of an aristocratic or even royal love affair with the affairs of government." "Courting relations," Altman writes, "are tied to government concerns."[71] At the center of the operetta's imaginary universe is thus typically the motif of the small kingdom or Old World principality, dominated by a castle or estate as the organizing center of narrative action: a place for the convening of lords, ladies, and their retainers, the castle-space provides the trappings for romantic and familial complications whose resolutions also serve to reinstate political order.[72] The film operetta immediately adopts this syntax, beginning with Warner Bros.'s *The Desert Song* and Paramount's Ernst Lubitsch-directed *The Love Parade* in 1929. The former, based on Sigmund Romberg's 1926 operetta, situates its *Sheik*-like romance against the backdrop of a rebel uprising against a French outpost in Morocco; the latter, Lubitsch's first sound film, ties the romance between the unmarried queen of Sylvania (Jeanette MacDonald) and a capricious count (Maurice Chevalier) to the restoration of the country's financial security. Yet the extent to which the operetta thus concentrates on adult themes of governance brought dilemmas for the Roach Studios' engagement with that tradition. True, the presence of clown or fool figures was a standardized trope of the operetta form, extending back to the style of opéra bouffe pioneered by Jacques Offenbach in the 1850s. But operetta's thematic organization also displaced those clowns from the kind of narrative pertinence that feature-length starring vehicles would seem to require.[73] The formal problem of the Laurel and Hardy operettas thus lay in a social and narrative conception that paradoxically both enabled and limited the duo's presence all at once.

These dilemmas were clear from the Roach Studios' very first operetta, *The Devil's Brother*, in which Laurel and Hardy appear as the comic henchmen from

Auber's original text, Giacomo and Beppo—here renamed Stanlio and Ollio. Although greatly expanded from the source, the henchmen's roles are none-theless caught between three nullifications that work to insulate their comedy from the social logic of the film's narrative. First, and most basically, Stanlio and Ollio are almost entirely excluded from meaningful plot agency. Predating the tradition of the operetta proper—as associated with Strauss, Lehár, and others—Auber's text belongs properly to the earlier lineage of the opéra comique and, accordingly, stakes its narrative on a somewhat different vision of social order, not the fantasy of a fairy-tale kingdom but rather the economic security of an emergent bourgeoi-sie (the same social vision, that is, as contemporaneous French stage melodrama, albeit here reworked as comedy); yet the narrative exclusion of the clown remains the same, for what is at stake is a social order that simply permits no space for clowning in its constitution.[74] The plot is thus configured around complications concerning property and real estate, of which Stanlio and Ollio have none. On the one hand, there is the romance plot: an innkeeper's daughter, Zerlina, loves a young soldier, Lorenzo, yet her father plans for her to marry into wealth so he will have enough money to keep his inn. On the other, there is the kind of risqué courtly intrigue characteristic of the filmed operettas of Lubitsch: a dashing bandit, Fra Diavolo, steals money and jewels from the young wife of a witless aristocrat, Lord Rocburg (the aristocratic couple played by Roach stalwarts Thelma Todd and Jimmy Finlayson). The two lines converge to resolve the innkeeper's economic situ-ation when Diavolo, unmasked as the thief, secretly gives some of the stolen loot to Lorenzo, who in turn uses it to save the innkeeper from having to sell his inn, so that it is not the riches of aristocracy that the narrative works to restore but the con-solidation of bourgeois economic security at the expense of bumbling aristocrats.

Where, then, are Laurel and Hardy in this story? Wandering peasants, they are tricked in their first scene into joining Fra Diavolo's team of bandits, but over the course of the film make only one meaningful contribution to the story's development—when a "spiffed" (drunk) Stanlio unwittingly reveals Diavolo's identity to Lorenzo. Instead, theirs is a life lived in the entr'acte, in the interstices of comic relief: in the time spent in bungling efforts to carry a sedan chair, or in repeat run-ins with an enraged bull, or in soporific drunkenness. "Interlude" was in fact a key term in the film's critical reception, where Laurel and Hardy's scenes were repeatedly described as "funny interludes," "pantomimic interludes," and the like.[75] What Stanlio and Ollio are subject to, in other words, is simply the other side of the same comic depreciation that renders aristocracy, in the form of Lord Rocburg, a figure of fun: laughter, in this narrative, radiates outward from the bourgeois middle to render all its outsiders fools, peasants and nobility alike.

It is in this sense that the duo's narrative marginalization is further crisscrossed by a second form of nullification, the motif of play, which defines the duo's comi-cality in terms of an infantilism entirely separated from adult concerns of mar-riage and property ownership. *The Devil's Brother* introduces this motif by having

FIGURE 30. "Stanlio" and "Ollio" in *The Devil's Brother* (1933). The caption on the back of this production still reads, "Stan Laurel does the 'finger-wiggle' for Oliver Hardy's benefit." Courtesy Academy of Motion Picture Arts and Sciences.

Stanlio repeatedly befuddle Ollio through the games of "Finger Wiggle"—which involves interlocking both hands by pivoting them around the middle fingers—and "Kneesy, Earsie, Nosey"—a challenge of coordination involving grabbing one's ears, knees, and nose with alternating right and left hands (fig. 30). (The reviewer for the *Herald-Tribune* described the latter game, which briefly became a children's fad, as "an exhibition of sheer idiocy . . . [that] convulses the audience.")[76] Subsequent operettas stuck to the formula: *The Bohemian Girl* featured Stan perplexing Ollie with a number of children's tricks, including the age-old "severed thumb" illusion, while *Babes in Toyland* launched another kiddie craze with its revival of "pee-wee," a game with a wooden stick and puck. (This time, the Roach Studios saw the merchandising opportunity and licensed the manufacture of a Laurel and Hardy pee-wee set.)[77]

Third, and most tellingly, Laurel and Hardy are also excluded from the world of song, which, typically of operatic forms, serves to galvanize the narrative's social universe, punctuating and defining its various political orderings. As philosopher Catherine Clément has observed, "In any opera there is what might be called a vocal backdrop . . . *a society of voices*. . . . The community is represented by crowds singing with a huge, vague voice."[78] From the outset of the film, song plays an essential part in introducing the narrative's various social groups—first the world of the bandits, next the world of nobility, third the world of the inn—before the clowns make their appearance as an eccentric fourth term, excluded from the film's musical organization. For the opening scene, we are introduced to Fra Diavolo singing his signature tune "On Yonder Rock Reclining" as he descends into the bandits' camp, greeted by a chorus of men and women with the refrain, "Diavolo, Diavolo, Diavolo!" The next scene is a flashback, as Diavolo regales his cohort with the story of his encounter with Lady Rocburg. The aristocratic world of courtly affairs and intrigue is here associated with the wistful barcarole with which Diavolo, disguised as a marquis, seduces Lady Rocburg into betraying the location of her jewels. Next, we are brought to the inn, where another choral refrain coalesces a third social world: a soldiers' drinking song ("Wine's the soldier's shield / In the tented field," etc.) introduces the community around the inn, where Lorenzo and Zerlina declare their love for one another.

It is only after these three scenes, paralleled in their use of song to define distinct social groupings, that Stanlio and Ollio are introduced, in what is, not coincidentally, the first of the film's scenes to lack song. Laurel and Hardy's exterior position vis-à-vis the narrative's broad social divisions—the world of the bandits versus that of nobility versus the bourgeois world of the inn—is thus defined as a specifically *musical* separation. There is no singing in Laurel and Hardy's introductory scene, in which they comically attempt to rob a woodcutter; instead, the script simply specifies "incidental noises to end of reel."[79] Indeed, there is no real singing for the pair at all in this film: clowning's close relation to musicality in the Roach shorts is, in this sense, inverted into a nonrelation to song in *The Devil's Brother*. Only at a much later point does Stanlio even attempt to sing, instructed by Fra Diavolo to use a snatch of melody as a secret signal in the plot to steal Rocburg's fortune. But Stanlio sings the wrong tune and is told to shut up: "If I hear a single note of that song again," Fra Diavolo yells, "I'll—I'll cut out your tongue!"

This use of song as a device for encoding a social structure from which the clowns are excluded carries over into *The Bohemian Girl*, which adapts Balfe's opera about a gypsy band and their kidnapping of an aristocrat's daughter. Once again, as with *The Devil's Brother*, the opening sequences use the "society of voices" device to organize and divide the different groups within the film's narrative: first, an extended sequence of gypsies singing and dancing ("Gypsy vagabonds are we / As free as anyone can be," etc.); next, a brief marching song as soldiers arrive at Count Arnheim's estate. Yet Stan and Ollie's place within this divided social world, and thus their relation to song, is again an ambivalent one, in which they serve primarily as intermediaries between the social poles of the narrative, fully belonging to neither. Nominally gypsies, they receive neither respect nor inclusion within their own community, which consigns to them the unwanted task of raising the kidnapped girl; nominally her deliverers, Stan and Ollie eventually restore the count's daughter to her rightful estate, only to be tortured, not rewarded. Nor, then, do they really belong to song, as again becomes comically evident in the one instance when Stanley *does* attempt a tune. Ollie hears a beautiful soprano voice singing outside his caravan, only for Stan to enter and reveal the voice as his (a woman's voice comically dubbed over). "Do you know you have a nice voice?" asks a surprised Ollie. "Oh, I had a much nicer voice until I ran a nail through it," is Stan's nonsensical reply. Later in the scene, Stan starts singing again—only this time he emits a deep bass for one verse, then a squeaky falsetto in the next (again overtly dubbed).[80]

Despite differences, then, in the political orders that *The Devil's Brother* and *The Bohemian Girl* enjoin—in the former, the economic security of a nascent bourgeoisie, in the latter, the familial lineage of aristocracy—neither is ultimately able to tolerate what literary theorist Mikhail Bakhtin described as the clown's "right to be 'other'" within their social worlds.[81] Neither permits the duo meaningful agency in the establishment of their respective orders, nor are they allowed a place once restoration is achieved: *The Devil's Brother* ends with Stan and Ollie being chased off by the bull one last time, *The Bohemian Girl* with them rescued from Count Arnheim's torture chamber. For the latter film's closing "gag," in fact, the duo appear as physical symbols of their own lack of fit, tortured into grossly misshapen form, Stanley squeezed short, Ollie stretched tall. (The *Variety* critic, with a view to the child audience, deemed this ending "downright unwholesome.")[82] In the meantime, the clown's power of play, as the right to refuse categories and expose masks, has been immunized as childish gaming, a kind of inconsequential fun to be enjoyed in the interstices of the adult world.

(An extended parenthesis is warranted here to touch briefly on a film that, while not one of Laurel and Hardy's feature-length operettas, nevertheless provides the most startling instance of these representational conundrums: *Bonnie Scotland* [1935]. Although lacking song, the film shares the operetta formula of linking romance to governmental concerns within an exotic "fairy-tale" space—here, the outpost of a Scottish regiment in the far reaches of imperial India. Borrowing liberally from the basic template of Sig Romberg's *The Desert Song*, the plot of *Bonnie Scotland* braids its romance around the regiment's efforts to quell an uprising of native forces, led by the evil Khan Mir Jutra. But the efforts of writers Frank Butler and Jeff Moffitt to find a place for Laurel and Hardy within that narrative broke down completely. In the preview cut, so much screen time was given over to the love story as to puzzle audiences and critics expecting a Laurel and Hardy feature. The "romance between a young soldier and the girl is developed almost entirely without the aid of the comedians," noted a *Variety* reviewer, while the critic for *Liberty* magazine similarly expressed disappointment at "prolonged scenes of a conventional and dully treated love story."[83] In response to the film's poor preview, Laurel elected to cut out much of the romance plot, reducing the original running time by some twenty minutes. In the process, however, he rendered the film's narrative incoherent. As the romantic lead William Janney remembered, "Stan cut out so much of it that when you'd see one scene, the scene before it was missing, so you'd wonder, 'What in the hell are they doing *that* for?'"[84])

It is thus telling that the one Laurel and Hardy musical feature to redress these dilemmas did so by bracketing off the concerns of the adult world altogether to situate its narrative fully within the realm of childhood fantasy. A free adaptation of Mother Goose nursery rhymes, Victor Herbert's 1903 operetta *Babes in Toyland* had already been eyed as a potential sound feature by RKO as early as 1930, when the company arranged to have Walt Disney produce the film as a two- or three-strip Technicolor animated feature for distribution through RKO exchanges; but when the expense of the project gave RKO cold feet, Hal Roach— who had fond memories of seeing the play as a young boy—stepped in to acquire the rights in November 1933.[85] *Babes in Toyland* became the most ambitious and expensive picture that his studio had ever attempted, its negative cost of $421,810 more than doubling the average of the firm's previous features (and requiring an exceptional $250,000 advance from Loew's-MGM).[86] Above all, though, *Babes in Toyland* represented a determined attempt to secure Roach's transition to features by targeting children as a pathway to the profitable family audience. The film was aggressively marketed in sentimental terms as a family picture for Christmas release. Promotional material explained how *Babes in Toyland*'s lavish budget had been spent to create "the average child's conception of fairy land," and the film was received in kind, with many critics commenting on its seasonal appeal.[87] "In England the 'Christmas Pantomime' for the kiddies is an ancient and honorable institution," the *New York American* reviewer observed. "[*Babes in Toyland*] provides something of the same sort for the youngsters here—the quiet little ones who sit up for Santa Claus."[88] "If Hal Roach aimed at the production of a purely juvenile picture to which children might conceivably drag their elders, he has succeeded in a measure beyond others who have sought to enter this realm," another critic noted, praising the film's "glamour of mysticism which marks juvenile literature."[89] (Far less maudlin, the reviewer for *Time* noted that "minors, for whom . . . [it] was presumably intended, are almost sure to like it," usefully adding, "A more important recommendation is the strong possibility that it will not bore, disgust or irritate their elders.")[90]

Cementing this sense of the film as a "special event" for children, *Babes in Toyland* was even given a well-publicized advance screening, on December 18, to an audience of some 175 disabled children at New York's Bellevue Hospital, where it was attended by two of the film's stars (Jean Darling and Johnny Downs, who appear in the film as Curly Locks and Little Boy Blue, respectively) and current Our Gang members Spanky McFarland and Scotty Beckett (fig. 31). The sentimental imperative that governed the film's marketing here crested in a staged event centered on the pathos of childhood dependency. As the *New York Times* reported, "There were tiny boys with arms in slings and casts or bandaged heads. There were children lying on stretchers, others seated in chairs, against which crutches were

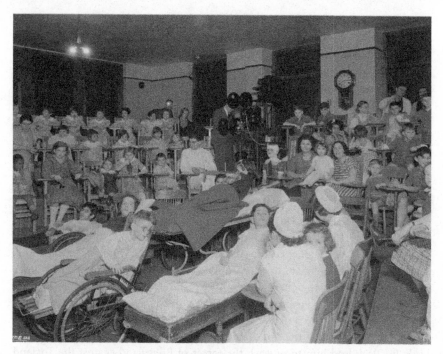

FIGURE 31. Special screening of *Babes in Toyland* (1934) to an audience of disabled children at New York's Bellevue Hospital. Courtesy Academy of Motion Picture Arts and Sciences.

propped. . . . As the children left the 'theatre' one little girl heaved a tremendous sigh. 'It's the best movie I ever saw,' she said."[91]

That *Babes in Toyland* was so explicitly marketed as a movie for children is the first clue here. In operettas like *The Devil's Brother* and *The Bohemian Girl*, we have seen, the clown's role was demoted to an interstitial space of childhood play, marginalized within plots revolving around adult concerns of marriage and political rule; in *Babes in Toyland*, however, the narrative world *is* that of childhood, insofar as it marks the exact adequation between narrative and nursery rhyme. To be childlike is thus not to be marginalized but to *belong*, for this is a world populated entirely by nursery rhyme characters. Little Bo-Beep, Tom, Tom the Piper's Son, the Three Little Pigs, the Cat and the Fiddle, the Little Old Woman Who Lived in a Shoe, and "Stannie Dum" and "Ollie Dee"—in the opening sequence, a pocket-sized Mother Goose introduces these dramatis personae one by one, turning the pages of a nursery book (a studio prop twice her size), each page bearing a matte live-action image of the character. Within this world, moreover, the themes of class power or political governance simply do not appear. Rather, to the extent to which this world is a contested space, it is so by forces whose possible social or political content is entirely absorbed into the received codes of childhood fable. As literary scholar Susan Stewart has argued, the fairy tale is typically a tale of miniature forces—a fable of fairies or dolls pitted against ugly dwarves, for instance—whose very reduction in scale creates an "other" time and space that displaces the world of lived, social reality. "The miniature," Stewart writes, "does not attach itself to lived historical time. . . . [It] skews the time and space relations of the everyday lifeworld, and as an object consumed, the miniature finds its 'use value' transformed into the infinite time of reverie."[92] Miniaturization is, indeed, the salient principle of Toyland's representation as a terrain for actualizing children's fantasy: to quote studio publicity, the set comprised "attractive little houses with candy canes forming a picket fence; the tiny homes of the Three Little Pigs built of bricks, straw and sticks; miniature wagons drawn by goats," and at its center, the boot-shaped home of the Little Old Woman Who Lived in a Shoe, here literalized.[93] The fantastic and diminutive scale of things is further established by a set of correspondences with familiar objects: the outsized book prop from the credit sequence, toy soldiers taller than the "human" protagonists, as well as "a number of giant trees . . . supplied by outside nurseries"—all coalesce in the impression of a toy-chest world whose precisely arranged detail rests on the exclusion of the grotesque, as embodied in the misshapen, imperfect form of the bogeymen, consigned to a nether realm.[94] Mediating the two spheres in *Babes in Toyland* is Silas Barnaby, the "meanest man in town," whose frustrated designs on Little Bo-Peep lead him to unleash the forces of Bogeyland against the Toyland residents. Only here that mediation is not between two halves of a social totality that must be arranged into a class order (as in the relation between bourgeoisie and aristocracy in *The Devil's Brother*) but rather of the illicit opening between two separate scales of fairy-tale reality—the miniaturized ideal, the monstrously grotesque—that must be hermetically sealed against one another.

It is here, in fact, that Laurel and Hardy finally find the narrative agency that in other features eludes them, playing kindly toymakers who become protectors of the Toyland harmony that Barnaby threatens. The boys' narrative contribution is at first unpropitious: they misunderstand Santa Claus's order for toy soldiers, building them six times their requested height. Yet the error nonetheless proves to be Toyland's salvation: in the face of Barnaby's bogeymen, Stannie and Ollie animate the mechanical soldiers and send them out to successfully ward off the attack (shot in a then-celebrated stop-motion sequence by Roy Seawright). They also turn the game of "pee-wee" into a strategy of offense, as Stan uses his expertise at hitting a puck to launch darts at the invading forces. The very childlike qualities that in other films inscribed Stan and Ollie's irreconcilable difference have here become the principle of their inclusion, at once the framework for their narrative agency and the means for the restoration of community (fig. 32). In contrast with other Laurel and Hardy features, critical opinion on *Babes in Toyland* thus celebrated the integration of plot and comedy: "Everything done has a direct bearing on the story, and all of the comedy gags are written about this thread of narrative."[95]

FIGURE 32. Stan and Ollie help Tom-Tom and Bo-Peep escape from Barnaby's lair, an example of the duo's narrative agency in *Babes in Toyland* (1934). Courtesy Academy of Motion Picture Arts and Sciences.

Importantly, the distinction dividing *Babes in Toyland* from Laurel and Hardy's other operettas also corresponds to a transformation in literary models, a switch from the European opéra comique / operetta tradition, premised on themes of social ordering and political governance, to the imprint of the fairy tale in its characteristically Americanized form as a vehicle for utopian imaginings. The "classical" fairy tales of Hans Christian Anderson and the brothers Grimm had conceived of narrative as a symbolic form for socializing children into standards of belonging, but later English-language writers like George Macdonald and L. Frank Baum had disrupted this normative function by envisioning fairy-tale utopias informed by more egalitarian and communal ideals.[96] Baum's legacy is an important one for *Babes in Toyland*. Not only was the success of his original *Wonderful Wizard of Oz* in 1901 the direct inspiration for Herbert's operetta (indeed, the same librettist, Glen MacDonough, was responsible both for Herbert's operetta and for the stage version of Baum's novel, which both opened in 1903 and which contain numerous structural parallels), but it is also likely that it was Baum's 1910 *Emerald City of Oz,* the sixth in his Oz series, that provided the template for Roach's narrative reworking of the Herbert musical.[97] The most explicitly utopian of all the Oz tales, *Emerald City* similarly sets the representation of communal harmony against the threat of invasion by grotesque creatures—in Baum's novel, the Nome king and his rapacious Whimsies, Growleywogs, and Phanfasms. What becomes structurally decisive in the Roach film, accordingly, is thus not the orchestration of a stratified social space, as in *The Devil's Brother* or *The Bohemian Girl*, but rather, as per Baum, the absolute contrast between fairy-tale utopia and the forces of its negation. There is thus a radical difference in the nature of the political orders that the Laurel and Hardy operettas enjoin; whereas their Viennese-style musicals work to establish a class-based ordering, be it bourgeois (*The Devil's Brother*) or aristocratic (*The Bohemian Girl*), *Babes in Toyland* envisions the realization of a communal utopia to which the label "political" can hardly be said to apply at all.

But the twist is this: that utopia exists only for childhood. As the lyrics to Herbert's "Toyland" remind us, first sung by Mother Goose as she introduces the film's characters:

> When you've grown up, my dears,
>   And are as old as I,
>   You'll often ponder on the years
>   That roll so swiftly by, my dears,
>   That roll so swiftly by,
>   And all the many lands
>   You will have journeyed through.
>   You'll often recall,
>   The best of all,
>   The land your childhood knew,
>   Your childhood.

The issue of the clown's narrative representability in operetta thus encounters both its resolution and, arguably, a dead end in the construction of an unreachable utopia whose forms are projected into an atavistic past. Only within the time-space of childhood fairy tale were Roach's filmmakers able to find a feature-length operetta format for fitting the clown to the role of narrative agent. But the limit of this solution lies in the fact that this utopic time-space is ultimately *presocial* in its orientation: if, for children, the fairy-tale space of Toyland is a realm for play antecedent to the experience of the social world, then for adults it is available only as an occasion for nostalgic retreat subsequent to that experience; for both, lived historical time is excluded.

*

Babes in Toyland was by no means the last attempt of Roach's filmmakers to find a working formula for Laurel and Hardy's musical features. The film was followed by two more musical features, the aforementioned *Bohemian Girl* and, in 1938, *Swiss Miss* (an original operetta, with songs composed by Phil Charig), both of which restore the duo to the interstitial roles of *The Devil's Brother*. (As Roach himself recalled of his original story idea for *Swiss Miss,* in which Laurel and Hardy play mousetrap salesmen who travel to a Swiss hotel, "What I was trying to do is make musicals where a second plot carried on, so that Laurel and Hardy didn't have be on all the time.")[98] But if it was not the last word, *Babes in Toyland* did provide the most definitive early example of a template that would become a minor trope of cinematic clowning in sound-era features: namely, the clown's conscription to fairy tale.[99] Despite some quibbles about the suitability of the bogeymen in a film intended for young children, reviewers were adamant in celebrating Roach's success in adapting slapstick comedy to the forms of children's literature. In the wake of the Payne Fund controversies over movies' influence on child development, *Babes in Toyland* was thus an object lesson in the value of fairy-tale-style family pictures and a model that producers like Disney and others would build upon. In terms of the specific trajectory of Laurel and Hardy's careers, *Babes in Toyland* established the form in which the team would be known and encountered by family audiences for years—indeed, decades—following. The film's substantial budget may ironically have ensured that it was the least profitable of all the early Laurel and Hardy features, with an initial net of only $13,853.24, but *Toyland* likely made up for this through many subsequent rereleases as a seasonal favorite.[100] Eventually, in 1948, Lippert Pictures acquired the film and reissued it, with a slightly shorter running time, as *March of the Wooden Soldiers,* the form in which it continues to screen to this day as a Thanksgiving and Christmas Day television special.[101] In terms of the broader arc of sound-era slapstick, the film helped cement a regular, if modest, trend in features that consigned the slapstick clown to a kind of no-man's land of childhood reverie—eccentric dancer Ray Bolger as the Scarecrow in *The Wizard of Oz* (1939), Abbott and Costello in *Jack and the Beanstalk* (1952), Jerry Lewis in *Cinderfella* (1960), or the Three Stooges in *Snow White and the Three Stooges* (1961).

There was nothing new in the clown's conscription to fairy-tale fantasy, and the earlier-quoted reviewer who linked *Babes in Toyland* to British pantomime was more perspicacious than he knew. The same historical process had occurred a century earlier on the British stage, where, by the 1850s, the harlequinade style of performance, in which clowns were the leading participants, had been marginalized in favor of child-oriented seasonal spectaculars and pantomimes based on nursery rhymes and moral fables.[102] The clowns were never excluded from these extravaganzas; rather they brought their clowning into them. The consequence, however, was in all instances a kind of circumscribing of comedy's carnivalesque potential as a site for the festive inversion of the status quo, a restriction of the clown's "right to be 'other'" to the realm of childhood. If the promise of comedy is akin to that of what Michel Foucault once termed "heterotopias"—those spaces within a given culture that function as "something like counter-sites, a kind of effectively enacted utopia in which the real sites, all the other real sites that can be found within the culture, are simultaneously represented, contested, and inverted"—then the heterotopia of fairy tale displaces that promise by twisting away from "real sites" to make itself absolutely anterior to them, the lost possibility of childhood rather than an alternative to the present.[103]

In the case of the Hal Roach Studios, I have been arguing, these processes can at least partly be explained through the studio's experiments with operetta as a musical and narrative framework for its Laurel and Hardy features; for, in the transition to operetta-style features, music ultimately operated as a principle of ordering that came at the cost of the clown's relation to lived social time. Whenever the studio operated outside this framework, its features were commonly dismissed by critics as padded-out two-reelers—as had been the case for Laurel and Hardy's debut six-reeler, *Pardon Us,* and as would also be true of subsequent nonmusical features like *Our Relations* ("Tedious after a few hearty reels," "They should know when to stop—and that is after the third reel") and *Block-Heads* (1938; "Should have been one of their usual 'shorts,'" "A lot of taffy stretched to the breaking point").[104] While the more prestigious format of operetta went some way to mollifying these complaints, it also tended to subtract from the currency of the team's comedy, now delegated to the interstices of a mythic political order configured through song (*The Devil's Brother, The Bohemian Girl*) or consigned to a realm of fairy tale accessible only to childhood (*Babes in Toyland*). What begins to emerge, in short, is a second mode of what the previous chapter discussed as slapstick's "social aging," only here linked less to a change in slapstick's audience (as in banalization) and more to textual strategies that served to withhold the clown from any relation with the social world. The next chapter extends this line of analysis by turning to the related issue of nostalgia and the emerging market for "old-time" comedy of the mid- to late 1930s.

5

"From the Archives of Keystone Memory"

Slapstick and Re-membrance at Columbia Pictures' Short-Subjects Department

The mood of retrospect seems indeed the soundest of possible instincts, fulfilling a purpose against which almost every large force in the country seemed to war upon, that to take root.
CONSTANCE ROURKE, *AMERICAN HUMOR* (1931)

If anybody else says it's like old times, I'll jump out the window.
BUSTER KEATON, IN *LIMELIGHT* (1952)

If one asks the naïve question "When was American film comedy's golden age?" one encounters the paradox that there has only ever seemed to be one answer: the silent era, specifically sometime between the ascent of Chaplin in the mid-1910s and the coming of sound. Often cited, James Agee's eloquent 1949 *Life* essay, "Comedy's Greatest Era," is a turning point in this regard, a nostalgic paean that, once and for all, elevated silent comedy as a symbol of the past glories of popular culture. "Anyone who has watched screen comedy over the past ten or fifteen years is bound to realize that it has quietly but steadily deteriorated. As for those happy atavists who remember silent comedy in its heyday and the bellylaughs and boffos that went with it, they have something close to an absolute standard by which to measure the deterioration."[1] It is a remarkable rewriting of slapstick comedy. Agee's essay was crucial in establishing Chaplin, Keaton, Lloyd, and Langdon as a kind of Mount Rushmore of comic achievement—the "four most eminent masters"— and it did so, we have seen, by appreciating slapstick in formal terms as an art of pantomime (see the introduction). No longer is the form criticized, as it once had been, for its vulgar intertexts in cheap amusements or its appeal to working-class "friends of burlesque."[2] Rather slapstick is reified outside of the network of social relations in which it had initially found meaning and celebrated for the abstract artistry of physical form, for the "beauties of comic motion which are hopelessly beyond reach of words."[3]

Of course, golden ages are always contingent constructions, invented, never discovered. I want to begin this chapter by tracing the outlines of a different slapstick nostalgia that emerged almost two decades prior to Agee's famous essay, more or less immediately following the transition to sound. The nostalgic idea of silent slapstick as a source of pantomimic beauty may have belonged to Agee (and later to Walter Kerr and others), but it certainly did not belong to the early 1930s, which cherished its own, more rambunctious memories of the form. Back then, it had been the rougher-edged style of early Mack Sennett shorts, not the refinements of later silent-era slapstick features, that stood for the genre's achievements; nearly all the silent clowns could be, and were, misremembered during this earlier period as roughhousing pie-throwers, not pantomimic artistes. "What might do this country a lot of good is a return engagement of those old custard pie throwing pictures that Chaplin used to be in," the *Motion Picture Herald* declared in 1932, in approval of one exhibitor's decision to run a reissue of Chaplin's actually quite un-custardy *A Dog's Life* (April 1918).[4] A taste for what was now described as "old-time" comedy spurred many distributors to reissue 1910s-vintage comedy shorts and many production companies to release pastiche recreations.[5] The early 1930s thus witnessed at least two efforts to bring back to the screen the "original" Keystone Cops, as in a reunion of sorts: the first, the comic melodrama *Stout Hearts and Willing Hands* (June 1931)—the debut entry in RKO's *Masquers Club* series of two-reel burlesques—brought together "the original Keystoners . . . Ford Sterling, Chester Conklin, Mack Swain, Hank Mann, Jimmy Finlayson and Clyde Cook"; the second, the Vitaphone short *Keystone Hotel* (August 1935), featured, to quote one ad, "the whole kaboodle of old-time favorites in a new 2-reel custard opera! . . . the Keystone gang in all their pie-eyed glory!"—again, Conklin, Mann, and Sterling, but also former Bathing Beauty Marie Prevost and cross-eyed comic Ben Turpin.[6] A similar move to tap the "old-time" comedy market came from the Hal Roach Studios, which, in 1932, launched a series of "Taxi Boy" comedies, planned as a throwback to the freewheeling, car-crashing comedies that Sennett had helped make popular. A letter from Roach studio executive Henry Ginsberg to MGM's New York distribution office indicates that the series was begun "because we need good slapstick comedy," and the films were promoted in kind. "The good old days of slapstick are back again!" declared an ad for the series in July 1932. "Isn't it the truth that short-reel comedies owe their original development to good, hearty *belly laughs* that rock an audience. No polite little laughs in this series. . . . Hal Roach estimates there will be 100,000 taxicabs ruined during the filming of this series. *That gives you a general idea!*"[7] Initial entries in the series constituted perhaps the most faithful continuation of the silent slapstick style outside of Chaplin's

1930s features: hired by Roach to helm the series, former Sennett director and stunt specialist Del Lord reintroduced the distinctive undercranking of silent-era comedy, shooting action sequences silent with sound effects and sparse dialogue dubbed in later.[8] The return of the "good old days" of slapstick was also seemingly announced that same year by the release of the Paramount feature *Million-Dollar Legs* (1932), a throwback farce starring W. C. Fields together with a cast of Sennett veterans—Turpin, Mann, Vernon Dent, even Andy Clyde—and direction from Sennett regular Eddie Cline, "the old maestro of slapstick days." "Good old-fashioned lowdown farce of the Sennett-Christie-Chaplin school," opined Leo Meehan in *Motion Picture Herald*.[9]

What does it mean that silent-style slapstick was so quickly revisited early in the 1930s in the form of what was called "old-time," "old-fashioned," or "good old-fashioned" comedy? This chapter seeks an understanding by looking in detail at the short-subject manufacturer that would come to be most closely allied with the emerging throwback market: Columbia Pictures' short-subject division. Famous today as the near quarter-century home of the Three Stooges (1934–1958), the Columbia short-comedy unit built its reputation as one of the industry's chief sustainers of earlier slapstick traditions. At a time when Sennett's studio was teetering on the verge of bankruptcy and Roach was wetting his feet in features, unit head Jules White relaunched the Columbia shorts department as a refuge of sorts for slapstick veterans who were now finding fewer places to work. Columbia sound engineer—and later shorts director—Ed Bernds recalled the situation: "Mack Sennett was about to close up shop; he hadn't been able to cope with the transition to sound, and Hal Roach, deciding that two-reelers had no future, was converting to feature films. Many talented people—writers, directors, and comics, refugees from the Mack Sennett and Hal Roach studios—were available," and many chose to continue working at Columbia's lot at Sunset and Gower, where they sustained the style of an earlier era in slapstick's history.[10] Viewed from this perspective, the Depression-era emergence of the "old-time" comedy market might be approached as an episode in the labor history of slapstick clowns, who flocked to Columbia in the face of a film industry that offered them sharply reduced employment opportunities.

But the reconstitution of slapstick in a throwback mode also speaks to a basic change in what might be thought of as slapstick's *social temporality*. Most theorists of comedy have assumed a kind of simultaneity linking comedic expression to social experience; that is, comedy has been thought to offer a symbolic articulation of social tensions and transformations contemporary with its expression. For anthropologist Mary Douglas, for example, social change coexists with all joking as the latter's precondition. "A joke is seen and allowed when it offers a symbolic pattern of a social pattern occurring *at the same time*," she writes.[11] But if this is the case, what sense can be made of the pleasures of a form of comedy that survived— was, indeed, commercially recycled beyond—the social context that produced it?

What was the appeal of a style of "old-time" slapstick that even one of Columbia's most accomplished short-subject directors, former Educational producer Jack White (Jules's brother), confessed to not finding funny because "they had done that twenty years before"?[12] The issue here would be uninteresting if it were simply a matter of old jokes. Also at stake, I will argue, is the way Depression-era American culture seemingly found nostalgic resonance in the populist comic styles of an earlier era and yet, in so doing, irrevocably transformed their legacy. As a live, contemporary form in the 1910s, Keystone-style slapstick had once constituted a class-coded carnivalizing of existing social relations, a distinctively plebeian gesture of social inversion; reappraised as throwback in the 1930s, slapstick would fall out of the present and into the flattened and undifferentiated past of "good old-fashioned" fun, where comedians and styles that were once quite incompatible (e.g., Chaplin and pie throwing) could now be conflated nostalgically with one another, consecrated as a tradition, and hence neutralized.

"A TWO-REEL OUT-AND-OUT SLAPSTICK MACK SENNETT-TYPE COMEDY": JULES WHITE AND SLAPSTICK REVIVALISM

The first task of an interpreter of the "old-time" slapstick trend is to situate it within what was a broader fabric of nostalgic sentiment during this period. The word "nostalgia" was first coined as a medical term in 1688 by the Swiss physician Johannes Hofer, from the combination of the Greek words *nostos* (return) and *algos* (sickness), to name the condition of homesickness he observed in young Swiss soldiers serving abroad. Two subsequent displacements would give the term its modern meaning in the passage through the late nineteenth and early twentieth centuries. First, the link with a lost place (*home*sickness) started to shade into the idea of an idyllic lost time; second, the term's clinical meaning in the context of psychiatry and military medicine was joined by a more everyday usage, the idea of nostalgia as something like a "mood" beginning to take over from its reference to a medical condition.[13] Indeed, despite sociologist Fred Davis's oft-quoted claim that nostalgia did not enter popular vocabulary until after the war—and despite theorist Fredric Jameson's influential association of nostalgia with the "cultural logic" of late capitalism—it is clear that the modern conceptualization of nostalgia in fact developed much earlier in twentieth-century America.[14] The emergence of a market for cultural nostalgia should be dated not to the 1950s, as Davis assumes, but to the 1930s: already by the beginning of that decade, the concept had entered common usage as a term of positive appraisal in journalistic criticism (to celebrate, say, the "note of poignant nostalgia" of a piece of music or the "pleasantly nostalgic" mood of a stage production, to quote from *New York Times* reviews).[15] More significantly, it is also during the 1930s that we see the first evidence of a self-conscious "decade nostalgia" set upon the styles of a particular past era—specifically, the "Gay Nineties." As film historian Christine Sprengler has noted, the late Victorian era "invaded the cultural consciousness and, in the early 1930s, provoked reflexive thoughts on nostalgia itself."[16] Sprengler cites a 1933 *New York Times* article, titled "Victorian Days That Beckon Us," in which journalist P. W. Wilson explained how "along with on-sweeping modernism is a looking backward, a nostalgia for a bygone age and, of all periods . . . it is the Victorian Age toward which longing eyes are more and more directed."[17] Throughout the prewar years, feature films from Mae West's *She Done Him Wrong* (1933) to the James Cagney–starring *The Strawberry Blonde* (1941) fed into the "Gay Nineties" nostalgia market; even short-subject series like Educational's *Do You Remember?* exploited the fad.[18] So prevalent was the trend in film as to draw complaining comment from *New York Times* critic Bosley Crowther, who critiqued the growing "tendency toward nostalgic retrospection." "We have no bone to pick with nostalgia and honest sentiment; they can make for strong dramatic effects," Crowther allowed, adding, "But there has been an apparent inclination to overdo the hearts-and-flowers . . . [which] strikes us as being a distinct and deplorable retrogression in film."[19]

But the "Nineties" was not the era's only privileged object of retrospect. It is remarkable, for instance, how quickly the legacy of silent cinema itself came to be treated as a fondly remembered, albeit lost object during these years. Previous scholarship has linked this nostalgia for the silent screen to the launching of the Museum of Modern Art's first circulating film programs, beginning in 1936, which provoked an enormous response in the popular press. Certainly, the work of Iris Barry's Film Library at MoMA, and its initial programs, "A Short Survey of the Film in America, 1895–1932" and "Some American Films, 1896–1934," cannot be underestimated in fueling a nascent public interest in earlier cinema.[20] Still such museological initiatives are best seen as symptoms, rather than sources, of this dawning sense of the medium's past: at almost exactly the same time as Barry's initiative, industry observers were pointing to a trend of remaking "well remembered films" from the silent era, including, for instance, RKO's remake of the Italian studio Ambrosio's landmark *The Last Days of Pompeii* (1908, remade in 1935) as well as Fox's version of D. W. Griffith's *Way Down East* (1919, remade in 1935) and a British version of the same director's *Broken Blossoms* (1921, remade in 1936), among others.[21] Virtually simultaneous, too, was the emergence of a proto-camp fascination with the outdated styles of early film: the same types of nickelodeon-era one-reelers that Barry sought to reclaim for serious-minded attention were already being reissued as comic relief, often with derisive commentary, in short-subject series such as Educational's *Great Hokum Mysteries* (1932–1933), Vitaphone's occasional *Movie Memories* (1933–1934), and the Stone Film Library's *Flicker Frolics* (1936, 1942).[22] Such mockery notwithstanding, Hollywood evidently came to realize that there was a growing market for its own past as the decade progressed. The close of the 1930s saw the release of Twentieth-Century Fox's *Hollywood Cavalcade* (1939), a "marathon romance" starring Don Ameche and Alice Faye, set against the backdrop of early Hollywood (discussed in more detail in this book's conclusion), as well as a *March of Time* special on the history of American cinema, compiled in collaboration with the Film Library of the Museum of Modern Art.[23] MGM announced a never-completed feature titled *Nickel Show,* which would have traced the "history of theaters from the old 'store show' days to the era of luxurious picture palaces," while, also in 1939, Columbia Pictures' New York chief, Jack Cohn, launched the "Picture Pioneers" club to bring together industry veterans for the purpose of "swapping reminiscences and promoting good fellowship."[24] Little wonder that one critic noted how Hollywood, "long adept in glorifying any subject or personality," had of late been "indulging in a little self-glorification" of its own past.[25]

Slapstick comedy, though, remained the thin end of this throwback wedge. It was unique in having entered a cycle of "old-time" reissues and pastiche recreations some three to four years prior to the broader market in film history of the mid- to late 1930s, and it was the *only* presound genre to be specifically singled out for such treatment (there was no comparable market for, say, "old-time" westerns). The speed with which slapstick became an object of nostalgia can most straightforwardly be explained through the transformations to the comedy field wrought since the coming of sound. "Nostalgia thrives . . . on the rude transitions rendered by history," wrote Fred Davis in his classic study, *Yearning for Yesterday* (1979), and the diffusion of sound was certainly rude enough for most slapstick filmmakers and their audiences.[26] Still, a more nuanced understanding would interpret such nostalgia not as a simple reflex of "rude transition" per se, but rather as a symptom of slapstick's changing position within the field of cultural production. We have seen (chapters 1 and 2) how the first wave of short-subject talkie comedies generated new hierarchical distinctions separating sophisticated metropolitan humor from what was now viewed as outdated "hokum" comedy. We have also seen (chapter 3) how slapstick's designation as "hokum" further implied a kind of slackening of the form's social currency, an inversion of the form's relation to modernity linked to its allegiance with the squarer sensibilities of the heartland. It is as though slapstick was becoming old thrice over: old in the basic sense of the advancing biological age of its practitioners; old in the sense of outdated, through the emergence of new comic styles; and old by association with marginalized taste cultures constituted by a conservative cultural remove.[27] Nostalgia, in this sense, stood for a kind of affective logic whereby the "old" could be reinvested as "old-time," outdatedness reclaimed as surplus value; it was, in other words, the very process through which slapstick, as a residual form, nonetheless sustained a lingering place within the mass cultural market.

*

It is within this context that Columbia's short-subjects unit first gained a reputation as a refuge for slapstick old-timers who found work there at the downside of their careers. Later in the unit's history, Jules White would be more willing to sign younger comedians like Johnny Downs or Sterling Holloway, but at the start, the unit's stars were a veritable parade of veterans: Charlie Murray (who turned sixty-two the year of his Columbia debut), Andy Clyde (forty-two), Harry Langdon (fifty), Charley Chase (forty-four), and Buster Keaton (forty-four) all starred or costarred in their own series; Polly Moran (fifty-three) and Slim Summerville (fifty-one) each were featured in Columbia shorts on two-picture deals; while Sennett veterans Bud Jamison (forty) and Vernon Dent (also forty) appeared in countless secondary roles. (Already by the mid-1930s, industry observers noted how Jules White was seemingly on a mission to corner "virtually all the comedians whose antics are suitable for short subjects.")[28] And then there were the behind-the-camera talents. Columbia gag writers Felix Adler and Arthur Ripley, directors Harry Edwards and Del Lord had all worked for Sennett in the 1920s; writer-director Clyde Bruckman had collaborated with many of the great silent comics, including Harry Langdon, Harold Lloyd, and Buster Keaton, with whom he codirected *The General* (1927); and former Educational comedy producer Jack White—Jules's brother—contributed directing duties at Columbia under his alimony-escaping pseudonym "Preston Black." Few comedy units were as well positioned to exploit the throwback market as Columbia's.

Still, silent comedy was never an undifferentiated field, and the throwback aspect of Columbia's product needs precise delineation. What is clear in this respect is that the formal achievements of 1920s-era slapstick counted for very little in shaping Columbia's house style, despite the presence in White's unit of some of the most notable comedians from that decade (Langdon, Chase, Keaton). The very qualities that have led later critics and historians to place a premium on the late silent era—the successful integration of slapstick with narrative values, the pantomimic virtuosity of the era's master clowns—all of this was occluded at Columbia in favor of the "roaring, destructive, careless energy" (the phrase is Gilbert Seldes's) of earlier, Keystone-style slapstick.[29] For example, Jules White's reliance on violent sight gags is notorious, nowhere more so than in the films he directed for the Three Stooges. But it is possible to see in this a return to the frenetic physicality of early 1910s "pie and chase" comedy, particularly in the tendency of the Stooges films to allow violent spectacle to derail narrative coherence and causality.[30] (Compare any of the Stooges shorts with a Sennett-directed early Keystone like *The Fatal Mallet* [June 1914], where a narrative of romantic rivalry serves simply as a springboard for an escalating sequence of excruciatingly violent knockabout. "The picture, purporting to be nothing other than a mélange of rough-house happenings, . . . proves that hitting people over the heads with bricks and mallets can sometimes be made amusing," commented the critic for *Moving Picture World*—words that could describe almost any of the Stooges films.)[31] At a time when other short-comedy units had abandoned the roughhouse style of 1910s comedy, White's department preserved a space throughout the 1930s and 1940s (and even into the 1950s) where such residual comic traditions could be sustained. "Columbia was a two-reel out-and-out slapstick Mack Sennett-type of comedy," commented Jules White in a late interview: "Mack Sennett's was the training ground supreme. That was the comedy college of all time. What we learned was how to steal from Mack Sennett . . . and rewrite it."[32] White also insisted on the importance of physical horseplay and roughhousing to his approach to comedy: "If we removed the knockabout aspect from these comedies it would have taken away their appeal. Their flavor would be gone. . . . It's like with westerns: take away the cowboy's six-shooter and there's no gunplay; without gunplay, it's a lousy western. The same applies to slapstick comedy: no slapstick, no laughs."[33]

Biography is relevant here, since Mack Sennett's Keystone studio was quite literally part of the landscape of Jules White's childhood. The Whites (originally Weiss) were Hungarian Orthodox Jews who had emigrated to the United States in 1904, eventually settling in Edendale, California, one of the early centers of moviemaking in the Los Angeles area. Home to the Keystone Film Company, Selig Polyscope, Bison, Broncho, and, briefly in 1915, Hal Roach's Rolin Film Company, among others, Edendale provided work out of school for the four White brothers (by seniority, Jack, Jules, Ben, and Sam), making childhood memories inseparable from recollections of early Los Angeles film culture, Keystone in particular. Jack remembered cutting his teeth as an office boy and kid actor at Keystone; brother Jules recalled a youthful prank that almost burned down the Mack Sennett studio; while Sam, the youngest, remembered selling newspapers to Sennett employees.[34] Jules's subsequent career path followed his older brother's, leading him to work at Educational, first as assistant film editor, then, by the mid-1920s, as a full-fledged director for the Mermaid and Cameo product lines. By the time of the transition to sound, the major studios' shift into short-film production had placed a premium on behind-the-camera organizational talent, and Jules found himself recruited to MGM to help organize its short department in 1929, then to do the same at Columbia. During his MGM stint, White directed several in Pete Smith's *MGM Oddities* series and, with director Zion Myers, created the *Dogville* comedies— spoofs of then-popular films featuring canine casts ("All-Barkie" comedies, as they were promoted), including titles like *The Dogway Melody* (December 1930), *So Quiet on the Canine Front* (January 1931), and so on. All-animal casts had previously appeared in silent-era comic series in the form of Hal Roach's *Dippy-Doo-Dads* comedies (1923–1924) but, in White's and Myers's hands, represented an ingenious solution to the challenge of sound filmmaking: toffee on the *Dogville* dogs' tongues made their mouths chomp repetitively, allowing for easy dubbing both in English and for foreign-language markets. Asked in later years to name the favorite of his short-subject series, White did not hesitate to answer "the dog things" (fig. 33).[35]

But it is fair to say that the major impact of White's approach to comedy rested not with masticating canines but with his subsequent work at Columbia Pictures. Founded in 1924 by Harry Cohn, Columbia had established itself by the beginning

FIGURE 33. Dogs in military training in So *Quiet on the Canine Front* (January 1931), one of White's early "All-Barkie" comedies for MGM.

of the sound era as one of the "little three" major studios, alongside Universal and United Artists—so-called because these companies did not have exhibition chains. Lacking a network of theaters, Cohn's organization obviously had little obligation to offer contracted exhibitors a full program, and the studio's involvement in short subjects had, until the early 1930s, focused primarily on the distribution of independently produced shorts. At the beginning of the decade, for instance, Columbia was handling Walt Disney's *Silly Symphonies* cartoons, Charles Mintz's *Krazy Kat* series, and Photocolor's *Color Sensation* musical comedies, although it also produced its own *Columbia-Victor Gems,* featuring vaudeville and musical stars, and the entertainment news weekly, *Talking Screen Snapshots.* Yet, if the lack of a theater chain inevitably placed short subjects at a low priority, Columbia's minor-league status also meant that there was little to lose by exploring genres that the vertically integrated studios were ignoring or downplaying. Slapstick shorts precisely fit the bill in this respect, and in the summer of 1932, Cohn signed the team of White and Myers to reorganize the studio's short department to emphasize cheaply produced live-action comedy: productions would be budgeted at bargain-basement levels of fifteen thousand dollars apiece, with shooting schedules of four to five days.[36] White later recalled his first encounter with Columbia's legendarily tyrannical head:

> I walked in[to Harry Cohn's office] and got just about abreast of the piano when Cohn says, "Hey, White. Can you make the funniest comedies in the picture business?"
>   I stopped cold and faced him. "Hey, Cohn. Can you make the best features in the picture business? Wait a minute, don't answer, I've seen them." I turned around and

started to leave. I got just about halfway to the door and he hollered at me, "Hey, where the hell you going, come back here." . . .

I said, "You still want to talk to me?"

"More than ever. I like a man that's got the guts to stand up and say what he's thinking."[37]

Despite White and Myers's previous experience together at MGM, things got off to a series of false starts at Columbia. Initial trade announcements of the new short-subject unit were followed just two weeks later by rumors of the unit's disbanding—apparently owing to a conflict over the famed vaudeville team Smith and Dale, assigned to White and Myers's unit by the New York office. (The trade press reported the conflict as a matter of scheduling; Jules White recalled it as a more sensitive issue: "I refused to work [with Smith and Dale]. I thought their type of Jewish comedy was demeaning and undignified.")[38] Reports of the unit's demise were in fact premature; even so, White temporarily quit in frustration the following year, having failed to produce a single film. Now under Myers's solo stewardship, the short-comedy division eventually sputtered into production on its first film on September 29, 1933, with its initial series falling in line with broader industry trends toward variety appeal.[39] First to commence release was the *Musical Novelties* series of musical comedy shorts, performed in rhyming dialogue under the direction of songwriter Archie Gottler (the sixth of these, *Woman Haters* [May 1934], was the first Columbia film to showcase the talents of the Three Stooges, there billed as "Howard, Fine and Howard"); next, a series reuniting the aging leads from Universal's *Cohens and Kellys* films, George Sidney and Charles Murray; and third, a line of domestic comedies seemingly patterned on Edgar Kennedy's *Average Man* shorts, featuring the bespectacled and befuddled Walter Catlett.[40] White's absence was nevertheless short lived; he returned to head the unit by the end of 1933—at Myers's request—and immediately set to work steering the short-subjects division firmly toward a throwback style of knockabout. The following year saw the launching of a number of full-bore slapstick series: the Three Stooges shorts (now in their own official series), along with the Andy Clyde (1934–1956) and Harry Langdon lines (1934–1945). Subsequent seasons rounded out the decade by adding more veteran performers: the pairing of former Keystoner and Fox Sunshine comedian Tom Kennedy with erstwhile Cameo comic Monte Collins (1935–1938); Swedish dialect comedian El Brendel (1936–1945); Charley Chase (1937–1940), fresh from being let go from Roach; and Buster Keaton (1939–1941), who, following his series at Educational, had been biding his time as an MGM gag writer. Reviewers were quick to recognize Columbia's throwback house style: thus, the Stooges' *Ants in the Pantry* (February 1936) was described as featuring "[gags] from the archives of Keystone memory," while the Collins and Kennedy short *New News* (April 1937) contained "gags of the kind familiar in Sennett's best years."[41] One exhibitor in Newport, Washington, waxed nostalgic in describing Andy Clyde's *He Done His Duty* (December 1937): "This was one of those old genuine comedies from those days when you say, 'Do you remember when?'"[42] Diverse as White's comedians were in style and approach, they would all be stamped with the trademark style of old-time roughhousing he helped nurture at Columbia.

"LANDING ON THE FLOOR WITH A THUD": THE AESTHETICS OF SLAPSTICK ATAVISM

A closer understanding of that atavistic trademark might usefully be approached by comparing the comedians at Columbia whose style had seemingly *least* to do with one another: the Three Stooges and Buster Keaton. Together, the Stooges and Keaton might be taken to designate almost opposed ends of the spectrum of styles within the American slapstick tradition, its highs and its lows. Whereas the Stooges' violently physical comedy has become for many a fitting emblem for slapstick's sound-era decline, Keaton's reputation—alongside Chaplin's, Lloyd's, and Langdon's—has come to define the pantomimic achievements of the previous decade, Agee's celebrated "greatest era." And whereas the Stooges have been seen as poster children for the throwback roughhousing favored by White, Keaton's work at Columbia is more usually dismissed as the unfortunately botched result of incompatible work methods and comedic philosophies—the disappointing offerings of a once great slapstick artist reduced to making assembly-line knockabout shorts, "alien to the true Keaton style."[43] What these comedians shared, however, was the experience of having to reshape their comic styles in interaction with work processes on the Columbia lot, with Jules White's tastes and creative agenda as well as those of his writers and directors. And although those experiences were very different, in both cases they resulted in a recursive or backward movement in the comedians' respective slapstick styles: a hopscotching over performative modes they had learned and mastered prior to their times at Columbia, and a restoration of comedic principles associated with an earlier moment in slapstick's development.

To begin with the Stooges (fig. 34): we have already had cause, in the first chapter, to discuss the team's earliest appearances in sound film—in particular, their five shorts with Ted Healy for MGM (1933–1934)—as examples of early sound comedy's indebtedness to Broadway's "cuckoo" vogue.[44] What changed in their comedy following their jump to Columbia Pictures was seemingly quite simple: no more Ted Healy. The Stooges had in fact long chafed at Healy's paternalism in controlling payments and contract negotiations and had even briefly quit the act as early as 1930, angered by his successful efforts to prevent Fox from granting the trio their own contracts. Despite reuniting in 1932, the Stooges quit Healy permanently when their MGM agreements came up for renewal on

FIGURE 34. Production still from the Three Stooges' *Healthy, Wealthy and Dumb* (May 1938). "HEALTHY, WEALTHY AND DUMB" © 1938, renewed 1966, Columbia Pictures Industries, Inc. All Rights Reserved. Courtesy of Columbia Pictures.

March 6, 1934, signing with Columbia on a one-picture deal for *The Woman Haters* less than two weeks later. (The separation proved to be a contentious one, with Healy filing an unsuccessful lawsuit seeking to prevent the use of the name "Stooges" in connection with the trio's subsequent pictures. Healy died, following a fight in a nightclub, in 1937.)[45] But with Healy more was lost than simply the boss's role in orchestrating the Stooges' mayhem. Also gone was the revue-style framework for which Healy had served as wisecracking master of ceremonies. Although *The Woman Haters* had featured rhyming dialogue set to music, the legacy of the Stooges' origins in musical revues in fact proved to be one of the major structuring absences of their post-MGM work: sans the revue-style framework of song and dance numbers, all that remained was the rowdiness with which the Stooges had previously punctuated their patter with Healy. Then as now, reviewers who described the Stooges' shtick at Columbia typically evoked its unbroken, relentless momentum. "Their routine," commented one, "is informal, inane and uninhibited. It's a series of eye-jabbing, head-thumping, nose-tweaking antics threaded on a string of rapid-fire chatter and embellished by double and triple takes"; "The Three Stooges," added another, "keep up a

running fire of gags and atrocious puns, interjecting the verbal onslaughts with
their familiar . . . antics, such as gouging the fat boy's eye, etc."[46] The social trajectory of the Stooges' comedy, in their passage first from stage to screen, and then
from MGM to Columbia, might thus be characterized as a gradual transition
from what had been the symbolically dominant pole (revue-style comedy) to a
symbolically dominated one (unadulterated physical slapstick), from a style of
comedy that enjoyed significant metropolitan cultural capital to one whose capital was in decline. There is, then, here a reverse development in comic style, a
move "back up the path" to an earlier—"lower"—moment in slapstick's development, prior to its incorporation as an element of spectacle in Broadway revues.

Sprung from this backward step, the Stooges' films thus became flag bearers for
the throwback comic style that became Columbia's, and Jules White's, hallmark.
They were only the third regular series launched under White's tenure (following the
Musical Novelties series and the George Sidney and Charlie Murray shorts), as well as
the unit's first truly slapstick comedies. Shortly after their debut in *The Woman Haters*
in May 1934, Columbia signed the trio for an additional seven shorts for the 1934–
1935 season at a thousand dollars per film (paid to the entire team, not individually),
to ascend to fifteen hundred dollars by their seventh.[47] Their third film for Columbia,
Men in Black (September 1934)—a spoof of the MGM/Clark Gable hospital drama
Men in White (1934)—brought White's fledgling unit unexpected endorsement in
the form of an Academy Award nomination, leading to further salary negotiations.
Jack Cohn, who supervised studio finances from New York, initially refused to raise
the team's salary to the suggested price of two thousand dollars a picture, but a letter
dated February 23, 1935, from studio manager William S. Holman convinced him
they were worth it. "In the first place," Holman reasoned,

> we feel quite certain that those boys will be able to and will sign almost immediately
> some place else. Accordingly, if we follow your suggestion we can be fairly certain
> that we will lose them entirely.
>
> In the second place if we were able to get the Stooges at $2,000 a picture, which
> we told you we thought we might do, they would not be as expensive for us as Andy
> Clyde now is and they would in reality cost us no more than Langdon does.
>
> The rate for Clyde is now $2,000 and his last two pictures go to $2,250, while
> Langdon is now at $1,500 with an option of two more pictures at $1,750. With both
> of these comedians we have to spend additional money for one or more supporting
> players, while with the Stooges we are getting three leads for one sum; therefore, as
> far as costs go, this price for them would not be out of line. Furthermore, we find
> it easier to get stories for these boys and also that their pictures on the whole are
> cheaper to make. Eliminating the football picture [*Three Little Pigskins*, December
> 1934] which went overboard largely because of weather conditions, their other three
> subjects have averaged under $12,000 a piece in cost, which is almost one thousand
> dollars less than the average for the other two-reel subjects. Also, the general feeling
> here is that their pictures are a little better and a little funnier than the general run
> of our two-reelers.[48]

So convinced was Jack Cohn, in fact, that he raised the offer to $2,500 per picture for eight shorts. The Stooges signed.

Such early success doubtless helped cement the Columbia short-subject unit's overall orientation toward live-action comedy, establishing the series as a boot camp of sorts for the unit's emerging style of intensively physical, throwback slapstick. It was director Del Lord who was perhaps the first fully to unleash the boys' destructive energies, regularly helming the Stooges' shorts between 1935 and 1945. As sound engineer (and future Stooges' director) Ed Bernds recalled: "When the Stooges came to Columbia, despite their year at M-G-M, they were still essentially vaudeville performers, using stand-up verbal routines; even their hitting and slapping routines were done vaudeville style." Under Lord, however, the Stooges' slapstick style became more expansive, replete with the stunt sequences in which the director had previously specialized at Sennett: "[Lord] brought them into the world of sight gags, special effects magic, and outrageous . . . stories"—as, for instance, in the barrel-chasing climax to *Three Little Beers* (November 1935), filmed on the streets of Edendale, Lord's old stomping ground.[49] A further turn to the destructive was added in their shorts under Jules White, who began directing them in 1939. Many of White's filmmakers later recalled how he pushed the Stooges to play up the direct, unmediated violence that was already part of their established stage repertoire.[50] Although typically White sought total control over the shorts he helmed, this was not the case with the Stooges; as he later recalled, he "permitted more ad-lib with the Stooges"—in particular, allowing Moe and Curly to develop elaborately violent tit-for-tats across lengthy, unbroken takes— and granted the team significant creative input into scripting.[51] Moe Howard later described the Stooges' role in scripting their shorts in a 1973 interview:

> Basically, we wrote about ninety per cent of the films we did on what they call a treatment basis. Our scripts were usually twenty-nine pages and we always wrote a nine-page treatment embodying what the background would be and different pieces of business and that kind of stuff. And then we'd turn it over to a screenplay writer who would put it in shooting script form. Then of course, we'd take the first draft of that and we'd take it back and put the speeches where they belong because you couldn't have Larry using any tough language to me because that's not the way we'd operate, you see? We'd change the dialogue and put the words where they belong and to whom they belong and then they'd take it back for a second and final draft, shooting script.[52]

The Stooges' working experience at Columbia was evidently characterized by a harmony of creative interests: on the one hand, a comedic trio whose style, absent the interplay with Ted Healy, was ripe for relaunching as "pure" physicality; on the other, a behind-the-camera team whose creative enthusiasms lay in the similarly retro direction of Sennett-style roughhousing. The same, though, can hardly be said of Buster Keaton, whose ten shorts at Columbia constitute perhaps the most critically reviled and neglected films in which he ever appeared. The critical consensus on these films has been damning: the ten Columbia shorts represent "the absolute bottom" of Keaton's career in films; they are "among the worst of his pictures" or "the worst comedies he ever appeared in."[53] Keaton himself shared the assessment, describing the films as "cheaters" ("movies thrown together as quickly and cheaply as possible") and claiming to have quit Columbia (in late 1941) when he "couldn't stomach turning out even one more crummy two-reeler."[54] Unlike other former silent stars at Columbia—Charley Chase, Harry Langdon—Keaton did not contribute much to the scripting of his shorts beyond preparing some routines in advance (he had, by contrast, coauthored all his silent comedies). And unlike the Stooges, Keaton's working relation with Jules White—who directed all but two of his Columbia shorts[55]—was less a matter of collaboration than of submission. Colleagues later recalled how White forced his silent-era stars to fit his conception of comedy by didactically instructing them on how scenes were to be performed. "Jules was an abortive ham," noted one colleague. "How could anyone have the audacity to show Buster Keaton or Harry Langdon how to do a scene? Imagine!"[56] White himself later explained this "audacity" in cruelly unsentimental terms: "They [White's lead comedians] all took my direction. They had to cause they were on their way down."[57] The gap separating Keaton's Columbia work from his achievements of the 1920s was thus cemented in his relation with an authoritative director-producer, Jules White, whose roughhouse inclinations harkened back to an earlier era. Ed Bernds recalled the disparity: "[Keaton] was good when he had a chance to be good. . . . But he was done in by . . . improper use of his talents. White's pace was frantic, but Buster's comedy required a more deliberate pace."[58] Financial limitations further compromised Keaton's ability to make these films his own, with budgets initially set at $15,750 per short.[59] As Keaton recalled:

> All of the energy and ingenuity of the director . . . [was] concentrated on the saving of money. . . . Several times I urged Harry Cohn, president of Columbia, to let me spend a little more time and money. I explained that on a larger budget I could turn out two-reelers that he could sell instead of giving them away as part of a package. Cohn, whose company was doing great without my suggestions, was not interested.[60]

Indeed, in Keaton's case, White not only contributed directing duties but was also responsible for the decision to pair Keaton with comedienne Elsie Ames in many of the shorts (fig. 35). Beginning with Keaton's fifth Columbia short, *The Taming of the Snood* (June 1940), Keaton's films were explicitly conceived and produced as costarring pictures in which Keaton would play foil to Ames, a dancer whom Columbia was building up as a brash comedienne somewhat in the Martha Raye mold (and who, not perhaps coincidentally, was at the time in an affair with White). Production files from this period are typically headed "Buster Keaton with Elsie Ames" (or some similar variant), and the films were even occasionally listed in the trade press as "Keaton & Ames" comedies.[61] The decision to partner

FIGURE 35. Production still of Buster Keaton and Elsie Ames in *The Taming of the Snood* (June 1940). "THE TAMING OF THE SNOOD" © 1940, renewed 1968, Columbia Pictures Industries, Inc. All Rights Reserved. Courtesy of Columbia Pictures.

VIDEO 5. Clip from *The Taming of the Snood*.

To watch this video, scan the QR code with your mobile device or visit DOI: https://doi.org/10.1525/luminos.28.7

Keaton and Ames seems to have been made following the production of *Nothing but Pleasure* (January 1940), the third in the series, in which Keaton had briefly appeared with Ames in a variant on one of the comedian's favorite routines: the famous scene from *Spite Marriage* (1929) in which he tries to put a drunk woman to bed (later performed with his wife Eleanor on tours in the 1940s and 1950s). Audience response to the sequence seems to have been positive—an Alfred, New York, exhibitor wrote *Motion Picture Herald* that "the part where Buster was trying to pick up the girl was especially good"—and the pairing quickly became policy.[62] For Keaton and Ames's first "official" outing as a duo in *The Taming of the Snood,* director Jules White and writers Clyde Bruckman and Ewart Adamson closely followed the template of *Nothing but Pleasure:* as in the earlier film, the pair's comic interactions are largely limited to a single, strenuously physical scene, again based on an earlier Keaton routine. The routine derived in this instance not from Keaton's previous filmography but from his family vaudeville act, the Three Keatons, for which he and his father Joe had performed a violent acrobatic act on, under, and around a table. (Joe Keaton was, in fact, often billed as "The Man with the Table.") In the film, Buster visits an apartment to deliver a hat; inside, he meets the maid, Hortense (Ames), who prances around, becomes drunk, and wrangles Buster onto a table. It is worth quoting sections from the shooting script's lengthy (three-page) outline of the ensuing action; here is the beginning:

41. *MED. SHOT—AT END OF TABLE*
Buster stands Hortense up then turns to pick up the hat. Hortense gets a bright idea [penciled: "tries to get up on table backward—jumps, falls"]. She lifts one foot, places it on the edge of the table, then reaches down, picks up the other foot to place it also on the end of the table, which naturally throws her. Just as it looks like she is going to hit the floor [this is penciled out], Buster whirls and catches her ["catches her" also penciled out]. He stands her up again beside the table. Her legs are wobbly and suddenly give way as she does the splits in a semi-collapse. Buster picks her up, straightens her out again and her legs start to go again. This time as she sinks, she reaches behind her to steady herself and grabs Buster by the back of the neck. As she continues to go down she pulls Buster over her shoulder. He lands on his head. For a second they look at each other. She laughs giddily, then gets to her feet. Buster is a trifle dazed. Hurriedly Hortense goes over to Buster and picks him up. Buster's legs are now wobbly and he starts to do splits. She straightens him up once more. [Last three sentences are penciled out and replaced with "he picks her up on his shoulder—she smashes his nose. He sits her on table"] . . .

And here are the elaborate acrobatics of the routine's finale:

55. MED. LONG SHOT
Buster is behind Hortense [him standing, her sitting on the table] and as she whirls around with her legs outstretched, pivoting on her fanny, her legs catch Buster in back of the knees, throwing his legs out from under him, and he lands on his fanny on the table, facing Hortense, the soles of his feet pressed against hers. Hortense's knees are in the air. She laughs giddily. She is having a swell time. Buster is beginning to burn up. Hortense, laughing hilariously, now suddenly straightens out her legs and the pressure shoves Buster off the table backwards. He lands on the floor on his neck. Slowly he starts to turn over, sits for a second angrily.

56. CLOSE SHOT—HORTENSE
She sees the lamp shade on the table and decides she will put it up on the chandelier. Picking it up she stands up.

57. LONG SHOT
Buster is slightly under the table where he sat up, and now as he turns over and starts to get up, his back catches the end part of the table, raising it. Hortense's (DOUBLE) feet go out from under her and she slides off the table on her fanny, landing on the floor with a thud on top of the hat. Buster hurries to her and picks her up.[63]

All this is remarkably close to the finished picture, providing unmistakable evidence that Keaton himself participated in the film's scenario preparation (who else would have known the routine?) (vid. 5). And, once again, it is this sequence that seems to have provided audiences with the film's standout scene: "Buster Keaton is never funny," wrote a dour exhibitor from Onalaska, Washington, but "this one is good. Credit to the 'drunk act' put on by the girl."[64]

Is it possible to suggest that, rather than detracting from Keaton's comic talents, as most Keaton scholars have argued, these moments with Ames instead supply the films' genuine significance for understanding the aesthetics of Columbia's "old-time" house style? Commentators often equate the Keaton-Ames partnership with the comedian's early 1930s work with Jimmy Durante at MGM—in *The Passionate Plumber* (1932), *Speak Easily* (1932), and *What! No Beer?* (1933)—but this is an observation that can be queried.[65] True, like Durante, Ames played wisecracking vulgarian to Keaton's mild-mannered simp, but where the Durante-Keaton contrast worked on a primarily *verbal* dimension (Durante was no knockabout comedian), the Ames-Keaton pairings are additionally distinguished by rambunctious and violently *physical* interaction. Frequently cast in roles of secondary narrative importance, Ames played characters who served simply to precipitate extended, elaborately choreographed slapstick sequences, not to advance the plot. If the gags in Keaton's earlier silent comedies are, as Noël Carroll observes, commonly task oriented, centered upon the display of Buster's "bodily intelligence" in overcoming physical obstacles, then slapstick in the Keaton-Ames films loses this purposive vector and is displayed purely for its own sake.[66] For instance, in *General Nuisance* (September 1941), the ninth in the series, Keaton plays Peter Hedley Lamar, Jr., a wealthy dandy who joins the army to impress a nurse (played by Dorothy Appleby; the plot is taken from Keaton's 1930 MGM feature *Doughboys*); what the film foregrounds, however, is a slapstick subplot in which Dorothy's gold-digging friend, Amabel (Ames), unsuccessfully attempts to woo Peter for herself. There is thus a discrepancy between the two women—Appleby's character serving as narrative "goal" and motivation versus Ames as agent of slapstick misdirection[67]—which generates one of the best comic sequences in both the film and the series: Keaton's clog dance with cuspidors on his feet. Entirely segregated from the Dorothy narrative, the sequence begins with Peter polishing cuspidors at the army camp; Amabel enters with an accordionist and serenades Peter, who takes up the melody and sings insults back at her ("Your teeth are not quite like the stars shining bright / But I bet fifty bucks they come out every night"). It is again worth quoting the script:

> At the end of the song Amabel grabs Peter and they go into a very nice Minuet. . . .
> Suddenly during the dance Amabel jerks Peter, who falls, Amabel dancing away from
> him and coming back again. . . . And now they go into a waltz. . . . In this routine Amabel
> and Peter waltz and the business of Amabel and Peter getting their arms all twisted up,
> the slapping routine, etc., until they separate, Peter flying across the room hitting the

wall. Amabel comes in and goes into a pose. Peter picks her up in a semi-ballet step. He lifts her up and as he lets her down, she stomps on his foot. Peter grabs his foot and begins to jump up and down. Amabel starts clapping in Russian tempo. . . . Amabel and Peter go into their Russian routine, ending with them doing leaps in opposite directions, CAMERA FOLLOWING PETER, who leaps into the cuspidors.

55. INSERT—PETER'S FEET
as they land in two of the cuspidors.

56. LONG SHOT
Peter now resumes the dance . . . as Amabel and he go into the clog waltz. During this an M.P. enters, comes down beside them and as Amabel and Peter separate in a series of turns, Peter gets the M.P. and waltzes with him while Amabel runs around him, trying to signal to Peter. Finally Peter sees that he has the M.P. and goofily dances away from the M.P., CAMERA PANNING WITH HIM. Peter hits the pile of cuspidors, ending up in a heap.

57. CLOSEUP—PETER
amidst the cuspidors.[68]

We can more closely approach an understanding of Columbia's house style by asking, Is there a "mechanics" to the comedy of these sequences? It has been a commonplace of Keaton criticism, for instance, that the gag structure of his silent films is in some sense "mechanical" or a symptom of technological modernity. Carroll, for instance, examines the style of Keaton's *The General* as embodying the "mind of an engineer," a pragmatic mentality focused on the "*mechanics* rather than the heroics of his physical encounters with things"; for Tom Gunning, meanwhile, Keaton's "operational aesthetic"—the display of mechanical processes in his gags—provides a paradigmatic instance of slapstick's intersection with the modern age.[69] But such judgments can be applied to the Columbia shorts only with significant qualification. For, as the showcase sequences with Ames illustrate, what counts in the Columbia films is less Buster's "mechanical" interaction *with objects* than an interaction *between characters mediated by objects*: Buster, Elsie, and a Murphy bed in *Nothing but Pleasure*; Buster, Elsie, and a table in *The Taming of the Snood*; Buster, Elsie, and a pile of cuspidors in *General Nuisance*.

The difference between tools and machines can clarify this distinction. As Lewis Mumford wrote in *Technics and Civilization*, "The essential distinction between a machine and a tool lies in the degree of independence in the operation from the skill and motive power of the operator: the tool lends itself to manipulation, the machine to automatic action."[70] But this is also one way of identifying the specificity of Columbia's hallmark comic style in relation to Keaton's silent-era accomplishments. Whereas Keaton's comedies in the 1920s incarnate a *mechanical* way of seeing—one preoccupied with the interaction of physical objects and human bodies in a more or less automatic process (e.g., the famous image from *Daydreams* [November 1922] in which Buster is trapped inside a boat's side paddle wheel, like a hamster on a treadmill)—then his Columbia shorts more commonly approach physical objects as *tools* to be manipulated in direct, frequently roughhouse interactions between characters. This is evident in its most basic form in moments of typical Jules White–style violence, when, for instance, Buster knocks out soldiers with logs of wood in a recurrent gag from the Civil War comedy *Mooching through Georgia* (August 1939) or uses a fork to fire mashed potatoes into the mouth of his ex-wife (Ames, of course) in *His Ex Marks the Spot* (December 1940). But it is also clear in the more choreographed sequences with Ames discussed above, in which physical objects function not as moving parts in an automatic or task-oriented process, but as inert material that either links the characters as a connective device (the table as both tool and topography of violence in *Taming of the Snood*) or serves as a direct extension of the bodily performance (the cuspidor clog dance in *General Nuisance*). Of course, historically, this use of a physical object as a tool—not a machine—for knockabout action has its paradigmatic instance in the "slap-stick" itself, a device from *commedia dell'arte* composed of two wooden slats and used to hit people (in Italian, a *battacio*). But this example again forcibly confirms how Columbia's house style led its starring comedians—Keaton no less than the Stooges—"back up the path" to an earlier modality of slapstick performance, one for which the mechanical dimension of physical comedy was not yet fully articulated, where the emphasis was instead on the interaction of human bodies, and where physical objects were employed straightforwardly as tools within that interaction (the hose of *L'Arroseur arrosé* [December 1895], the mallet of *The Fatal Mallet*, the supposedly ubiquitous pies and banana peels).[71] The distance that divides Keaton's Columbia shorts from his silent-era work is thus measured by a shifting modalization in the performer's comic relation to the material world of objects: no longer an assimilation of the body to a mechanical environment, but an appropriation of physical objects to human relations, even of the most brutal kind.

*

According to Pierre Bourdieu, it is the destiny of all cultural forms that age to see themselves "thrown *outside* history or to 'pass into history,' into the eternal present of consecrated *culture*." Historical variation within any given tradition is flattened into an undifferentiated field in which trends and schools once considered incompatible are barely even recognized as distinct; appreciation of that tradition instead becomes governed by what Bourdieu calls "transcendent and eternal norms," a unifying mode of perception that subsumes the dialectics of historical difference to a single static sameness.[72] The taste for "old-time" slapstick was just such a sensibility, and over the course of the 1930s, it dragged the particularities of distinct slapstick styles from the silent era into the gravitational field of Sennett-style comedy, as though all of the great slapstick stars had once desported themselves in a world of brick throwing and crazy chases. A Chaplin film like *A Dog's Life*, we have seen, could thus be thought of as a "custard pie throwing" picture. Keaton himself was often mistakenly thought of during this period as a veteran pie thrower—according to one report, the "Carl Hubbell of cinema pie-pitching."[73] (A player for the New York Giants, Hubbell set a major league record of twenty-four consecutive wins by a pitcher in 1936 and 1937.) Only a mode of perception for which slapstick had been consecrated to a flattened past could celebrate, for example, Harry Langdon, in Columbia's *Sue My Lawyer* (September 1938), as a "past master of . . . old-fashioned slapstick," just as only a mode of appreciation that blurred out historical variation under the blanket designator of old-time comedy could describe the style of Keaton's Columbia performances as being "in the manner to which, lo these many years, he has accustomed us."[74] The process is akin to what sociologist Barbara Myerhoff describes as "re-membering," the way in which nostalgia often seeks a "reaggregation of [a group's] members," an imaginary configuration of individuals who come to be associated in popular memory with a given historical event, process, or, in the case of silent-era slapstick, cultural tradition.[75] It was symptomatic, then, that during the 1930s Keaton was erroneously re-membered as a former graduate of the Keystone Film Company—a misperception that seems to have begun in journalist Gene Fowler's myth-making and error-ridden "biography" of Mack Sennett, *Father Goose*, published in 1934, and that, by the end of the decade was treated as more or less established fact.[76] For a cameo as himself in Fox's *Hollywood Cavalcade*, for instance, Keaton played a veteran of the swaddling days of Keystone-style slapstick, supposedly the first comedian ever to throw a custard pie (something he had *never* done in his silent days). The error was further perpetuated in trade press accounts, even in an educational supplement of *Photoplay Studies* devoted to the "real" history behind *Hollywood Cavalcade*, wherein it was baldly claimed that Mack Sennett had "invented" the "sad-eyed comedian."[77] Keaton's Columbia films might, in this sense, be thought of as the aesthetic corollary of this process of re-membering—a series of shorts that aligned Keaton with a performative style alien to his own silent films but perfectly fitting the dominant perception in the 1930s of what silent slapstick had been.

So it should come as no surprise that appreciation of Columbia's short-subject series was primarily oriented toward the pleasures of nostalgia. Even a cursory survey of *Motion Picture Herald*'s "What the Picture Did for Me" column confirms that Keaton's Columbia shorts attracted audiences precisely for their nostalgic appeal. Theater owner Pearce Parkhurst (Beverly, Massachusetts) informed readers that "this *old timer* [Keaton] still means something at the box office." W. Varick Nevins III, a theater owner from Alfred, New York, agreed: *Nothing but Pleasure*, he wrote, proved that "the *old time* comedians still have something the new ones don't." A somewhat less favorably inclined exhibitor, from Hay Springs, Nebraska, was surprised that audiences still found Keaton funny but conceded that *She's Oil Mine* (November 1941) had nonetheless been a success: "Just why an audience would go for this *old time* slapstick is more than we could figure out, but it seemed to bring out a lot of laughs."[78] It is telling that, of over two dozen exhibitors' reports on Keaton's Columbia shorts, only one expressed a totally unfavorable response (to *Mooching through Georgia*), and, furthermore, that that report came from the only big city exhibitor to comment on the series (the manager of Chicago's Plaza Theater): "OK for the children. A bit too silly for the adults. We received a big kick out of just this sort of a comedy fifteen years ago."[79] Regional distinctions separating small-town from metropolitan cultures were thus buttressed by a kind of temporality of taste; for, as the "What the Picture Did for Me" column shows, all the praise for Keaton's Columbia shorts came from small-town and rural exhibitors who valued their atavistic style as "old-time," while the only criticism came from a metropolitan theater owner who disparaged such atavism as outdated ("fifteen years ago").

A similar pattern is clear in relation to the Stooges, whose marketability was likewise defined in terms of the throwback tastes of small-town and neighborhood moviegoers to whom Columbia, lacking its own first-run houses, had long targeted its output. Of a film like the Stooges' *Hoi Polloi* (August 1935), the manager of the Niles Theatre in Animosa, Iowa, thus described the trio's comic style as "knock down, drag-'em-out" hokum, adding that "they [i.e., audiences] love it in the small towns on Saturday night," while an exhibitor in Lincoln, Kansas, praised the Stooges for finding "more ways to make people laugh than a farmer has coming to town."[80] As the Iowa exhibitor's comment indicates, small-town theater owners commonly arranged weekly schedules so that Stooge comedies played on Fridays and Saturdays, when downtown was crowded with rural folk, and they encouraged others to do the same. "By all means play all of the Stooges comedies," one Minnesota exhibitor recommended to his peers. "We play them on weekends on Friday and Saturday." The manager of the Palace Theatre in Exira, Indiana, agreed, noting: "When better comedies are made, Columbia and the Stooges will make them. . . . Play them on the week-ends and watch your audience really get a good laugh."[81] Children, in particular, responded favorably to the Stooges' shorts, making the pictures a particular boon for neighborhood exhibitors dependent on the family audience. The dialogue-heavy style of sophisticated talking shorts may well have alienated the kiddie trade, as historian Thomas Doherty suggests; but Stooge-style physicality was evidently an easier sell.[82] "Hard to get the kids out as they want to see [the Stooges] two or three times," commented E. F. Ingram of the Ashland Theatre in Alabama, an opinion with which many other exhibitors agreed.[83] "This trio of comedians are . . . nine-tenths of my box office appeal for the kids," noted the manager of the Gem Theatre in Logan, Utah. "You can have your Marx Brothers and your Laurel and Hardy, but give me just one feature-length Stooge picture and I'll be out of the red for a good while."[84]

What should hold our attention here is how Columbia's slapstick stars were only ever perceived to appeal to one side of a series of assumed cultural schisms: to children more than adults, to small-town moviegoers more than urban sophisticates. There was, then, a kind of structural homology mapping the social characteristics of Columbia's brand of throwback comedy onto those of their Depression-era fans, for both were, in a profound sense, *untimely*. Practitioners of a style of slapstick culturally situated in the past, Columbia stars like the Stooges and Buster Keaton became the objects of enthusiasms that were similarly marked by the "untimely" social positions of their users—whether children (defined through their presocial sensibilities) or heartland filmgoers (associated with the ambivalences of "provincial modernity"). Old-time comedy, as practiced at Columbia, thus marked a kind of temporal displacement linking the social position of its audience to a throwback cultural form. The next section considers how such untimeliness was not simply an issue of comic style but also governed Columbia slapstick's relation to the historical situation and populist ideologies of New Deal–era America.

"AN ARENA OF LIFE WHICH WE DON'T UNDERSTAND": THE THREE STOOGES AND NEW DEAL–ERA POPULISM

The Depression, no historian would disagree, brought the specter of the disenfranchised to the very center of American politics. It would thus seem an era suited to an earlier style of slapstick comedy, whose typology of clowns had long gravitated toward the marginal classes: to comic hoboes, befuddled immigrants, and the like. The destitute tramps that Keystone had put on screen in the 1910s, for instance, continued to hold sway in the political and social realities of the early 1930s, when unemployment reached 25 percent of the workforce while others received only a tiny number of hours of work per week. This, one might have expected, surely yielded opportunities for slapstick representation.[85] Yet even in the populist efflorescence of that decade, when so many agitated for an engaged popular art, slapstick remained strangely out of step with the times. Slapstick's aging had the effect of altering what I described earlier as its "social temporality," its ability to engage the social and cultural forms of the present as comic material. Old gags and formulas could only with difficulty be adapted to the changing ideological forms of American cultural life in the 1930s.

The films of the Three Stooges suggest this quite well. Few of the era's comic teams belonged so directly to slapstick's typology of disenfranchisement as the Stooges, who came to film already familiar to vaudeville audiences and theatergoers under a variety of lumpen designators ("Three Lost Souls," "Ted Healy's Racketeers," etc.). Early on, in fact, Columbia's screenwriters evidently sought to place the team in situations engaging the hardships of Depression-era life—for instance, *Three Little Pigskins*, which starts with the trio panhandling for cash; *Pop Goes the Easel* (March 1935), where they beg passersby for work; and *Movie Maniacs* (February 1936), in which the team is introduced riding the rails. There were also examples of direct political intertextuality in films like *Cash and Carry* (September 1937), which ends with the Stooges receiving a pardon from FDR, or *Healthy, Wealthy, and Dumb* (May 1938), whose plot satirizes New Deal–era tax laws.[86] Yet for every Stooge film that made a direct contemporary reference, there were far more that retreated into the abstractions of genre parody and costume farce. A slippage from contemporary settings to the occasional period comedy occurred very early in the Stooges' Columbia career, far more so than in any comparable comedy series from the period: already by their fifth, *Horses' Collars* (January 1935), the Stooges were branching out into spoof westerns; their sixth, *Restless Knights* (February 1935), was a medieval comedy; and their eighth, *Uncivil Warriors* (April 1935), Civil War slapstick.[87] All told, around a quarter of the forty-three Stooges shorts released in the 1930s placed the team at some kind of exoticizing distance, whether temporal, in, for instance, medieval comedies, or geographic, as in the Egypt-themed horror slapstick, *We Want Our Mummy* (February 1939); in such instances, there was a shading from the clown as figure of social disequilibrium toward disorder of a more abstract or even formal type, genre parody here eclipsing social content in seeming anticipation of the 1940s features of Abbott and Costello. Along similar lines, fan magazine and trade press profiles, although few and far between during their years at Columbia, painted the Stooges not as social satirists but as clownish figures of a quasi-mythic past. In one such, a 1938 interview in the *New York Daily Mirror*, the Stooges were sarcastically presented as upholders of the antiquated legacy of antebellum southern gentility: "Gentility and politeness were supposed to have died with the old South after America's Civil War," Moe Howard there claimed. "My brother, my partner and I simply felt that 'Bleak House' should not have entombed the only 'model of deportment,' so we devoted ourselves, all three of ourselves, to the cause of greater gentility."[88] Such evocations of timelessness would only be invoked more seriously in later years: "Like 'Old Man River,'" commented the *Los Angeles Daily News* in the early 1950s, "those wacky purveyors of slapstick, the Three Stooges, 'just keep rollin' along.'"[89]

Yet the question of social temporality perhaps comes into sharper focus if we approach the Stooges' films not only in terms of their settings and contemporary references, but also in relation to the comic *situations* that provided the plot frameworks for their comedy. This is because, like all narrative forms, conventionalized comic situations inevitably bear the traces of the social contexts—the attitudes, assumptions, and worldviews—within which they were originally generated. It thus becomes possible to read even the most rudimentary plot device as a condensed distillation of specific forms and moments of social experience—akin, in this respect, to what Fredric Jameson has dubbed "ideologemes," those zero-degree narrative conventions and clichéd plot devices that survive as fragments of former ideologies.[90] The slapstick tradition was, of course, replete with inherited devices, so much so that many acquired trade nicknames: the "bomb under the bed" plot (premised on alternating scenes between an unaware comedian and a situation of growing peril) or the "ghost in the pawnshop" (in which a clown is stalked by a menacing figure, mistakenly assumed to be a ghost).[91] Bears, lions, and gorillas on the loose; tramps who enter society; marital infidelities in adjacent hotel rooms; visits from decorators/builders/piano deliverers; millionaire benefactors revealed as escaped lunatics: such comic devices circulated so often that it is easy to mistake them as universal rather than historical forms that, in many instances, outlasted the social contexts they originally symbolized.

A case in point would be the basic master plot of so many of the Stooges' two-reelers: the disruptive entry of the lumpen onto the world of the elite. As Moe Howard himself glossed the formula, "We subtly, the three of us, always go into an area of life which we don't understand. If we're going to go into society, the picture opens on us as garbage collectors."[92] Time and again, regular Columbia writers like Adamson, Adler, Bruckman, and Elwood Ullman constructed narratives requiring that the Stooges transgress the boundaries dividing the dispossessed from the world of culture and privilege: they are, alternately, carpenters who inherit a Fifth Avenue dress salon and stage a fashion show (*Slippery Silks* [December 1936]), insect exterminators hired as society escorts (*Termites of 1938* [January 1938]), or telephone repairmen mistakenly employed as psychiatrists for a thrill-seeking society woman (*Three Sappy People* [December 1939]). Such situations were hardly original to the Stooges. As a basic slapstick situation, this "intrusion plot" gave expression to a geography of class segregation that had long served as a comedic imaginary for articulating workers' fantasies of revenge against their own exclusion. Born of the widening social contrasts of the nineteenth century, the trope had first entered popular narrative not as material for comedy but as a device of working-class dime novels like George Lippard's *New York: Its Upper Ten and Lower Million* (1853) or Frederick Whittaker's *Nemo, King of Tramps* (1881), which structured social space as a polarized landscape of privilege and destitution. But it also supplied a framework that proved adaptable to turn-of-the-century comedy, both on the vaudeville stage and, subsequently, in early cinema, allowing for numerous comic interactions between its two contrasting poles: on the one hand, workers, hoboes, misfit immigrants; on the other, millionaires, society matrons, and other "respectable" types. Nowhere, in fact, was this more pronounced than at Jules White's beloved Keystone Film Company, whose films of the mid-1910s developed an entire slapstick topography in these terms, consisting of mansions and street corners, ballrooms and saloons, swanky restaurants and hash houses—precisely the topography that White's Stooges would come to inhabit twenty years later.[93]

If this can rightly be described as a populist motif, then it was so crucially as a class-based, *plebeian* populism—that is, an allegorical and affective figure for articulating workers' experiences of disenfranchisement in a class society and the fantasies such disenfranchisement spawned. (Literary historian Michael Denning, in discussing nineteenth-century dime-novel fiction, describes these fictive figures of social cleavage as the "dream-work of the social.")[94] Doubtless the theme remained politically and socially symbolic following the 1929 economic crash; still, its derivation belonged to the more violent class antagonisms of an earlier period, to the decades that saw the earlier Depression of 1873 and the spread of poverty, unemployment, riots, and worker militancy—the Great Railroad Strike, the Haymarket bombings—in the years that followed.

By the time of Roosevelt's New Deal, however, this class-based populism had been complicated by the emergence of a more inclusive rhetoric. "During the period from 1935 until the end of World War II," historian Warren Susman writes, "there was one phrase, one sentiment, one special call on the emotions that appeared everywhere in America's popular language: the people."[95] Thus, whereas the populism of the earlier period was founded on a principle of division—the dispossessed versus the powerful—the New Deal spurred the emergence of a mainstream liberalism that, as we have earlier seen (chapter 2), tended to empty divisions of ethnicity, race, and class into the imagined civic solidarity of the "common man." A well-known example of this shift is provided in Kenneth Burke's controversial suggestion at the 1935 American Writer's Congress to shift the language of the left from "the worker" to "the people," arguing that "the symbol of 'the people' . . . point[s] more definitely in the direction of unity."[96] But the rhetoric was hardly limited to literary debates. Susman continues: "Thus Carl Sandburg gave us *The People, Yes;* the WPA projects offered Art for the People and The People's Theater; Frank Capra provided a series of enormously successful populist films in praise of 'the little man'; and John Ford, most significantly in a number of films made near the end of the 1930s, rewrote American history in mythic and populist terms."[97]

To insist on the prevalence of "the people" as an ideological motif is clearly not to suggest that the politics of class conflict were somehow dissolved under Roosevelt's New Deal. In an important corrective to Susman's reading, Michael Denning has warned against confusing the era's populist rhetoric with its populist practice: the rhetoric may have been inclusive, but politics on the left remained class-based and labor-oriented.[98] Still, granted the distinction between rhetoric and reality, there remains little question that a unifying mythos of "the people" *did* significantly influence the dominant cultural forms during this period, and what that mythos spurred, in part, was a new kind of "folk nationalism" predicated less on class militancy than on utopian themes of collaborative solidarity and cooperative aid. The turn to these themes remained open to a variety of inflections: great ideological fissures existed between conservatives who viewed the burden of mutuality primarily as a matter for individual philanthropy versus New Dealers who advocated federal action. Most mainstream Hollywood output followed more conservative winds in endorsing only the former, in sentimental narratives dramatizing purely individualized responses to social and economic inequity. This was the era, for example, in which charity was exemplified in Hollywood features in Shirley Temple's unstinting capacity to soften the hearts of the wealthy to bring their aid to the disenfranchised, and solidarity in Will Rogers's plain-spoken, intuitive intercessions against prejudice and injustice.[99] Laurel and Hardy meanwhile became icons of Depression-era companionship whose star personas split the difference between cooperative ideologies of assistance and individualist notions of up-by-the-bootstraps endeavor: on screen as in their reported off-camera lives, they helped (or tried to help) *each other* without depending on the help *of others*. Describing the career paths that led to their success, one MGM pressbook thus explained that "it was a long grind, they knew of no one who could pull a string here or a string there. Their success is entirely due to their own efforts," while in the next breath insisting that "Laurel and Hardy Success Due to [Their] Co-operation."[100] While their films—both features and shorts—explored such mutuality as comedy, with its attendant frustrations and squabbling, their off-screen personas embodied mutuality as a fully perfected ideal. "It is an actual fact and matter of record that Stan and Oliver have not had a single disagreement" was the unlikely claim of a puff-piece for their 1936 feature *Our Relations*.[101] "I don't know offhand any two actors who share equal starring honors with as much grace and apparent unselfishness as do Laurel and Hardy," Roach director Paul Parrott was elsewhere quoted as claiming.[102]

It barely needs saying that no such companionate depictions were ever given of the Stooges, whose press coverage instead emphasized how their hostile on-screen interactions carried over into their actual lives. "The private life of the Three Stooges is a subject we've wondered about for years," began a 1938 profile in the *New York Times*. "The first question we asked, when we finally had got two of the Stooges together, was pretty obvious: 'Don't you boys ever get on each other's nerves?' Moe . . . promptly answered: 'Yes.'"[103] Sentimental New Deal–era themes gained little purchase, apparently, from the Stooges' rowdy plebeianism; still, much of the interest of the team's films from this period lies in the way they occasionally registered the tension between these ideological frameworks, inventing provisional resolutions to the gap between them. These two populist paradigms, the sentimental-inclusive versus the plebeian-confrontational, which from the standpoint of changing class ideologies might be seen as historically distinct, could nonetheless be elided at the level of slapstick representation. In the classic, Keystone-style version of the intrusion plot, for instance, the lumpen clown creates a disorder that aggravates and exasperates, leaving the privileged classes to look on aghast at the destruction, but in a handful of Stooge films from the 1930s, the team creates a disorder that *binds*, unexpectedly uniting the opposed classes in a carnivalesque spirit that belies the rigidity of social rituals and distinctions. That is, there is confrontation, but a confrontation that generates a peculiarly inclusive disorder. In two instances, *Pop Goes the Easel* and *Three Sappy People*, the Stooges' irruption precipitates a vortex of escalating slapstick energies that ends up dragging the privileged willingly into its orbit. Thus pretentious artists acquit themselves with gusto in a clay fight with the Stooges in *Pop Goes the Easel*; society dinner guests answer the Stooges' pie-flinging antics with yet more pie flinging in the climax of *Three Sappy People*: in such instances, all hierarchies, all differences are erased in a shared spirit of enjoyable chaos. *Hoi Polloi* meanwhile adapts the intrusion plot into a kind of Stooge version of *Pygmalion*. Convinced that environment, not heredity, shapes social behavior, Professor Richmond (Harry Holman) bets a colleague he can take three trash collectors (the Stooges) and turn them into gentlemen. After much training, the boys are brought to a society event where they prove the professor's thesis, albeit by inverse: despite their best efforts at acting dignified, the Stooges' presence at the party has the consequence of bringing everyone down to *their* level. Stooge-style roughhousing spreads like a virus, from guest to guest. Society matrons start telling their husbands to "spread out," respectable men begin jabbing one another in the eye; by the end, it is only the Stooges who bear themselves with some semblance of comportment as the party degenerates into a free-for-all. From the final pages of the script:

FULL SHOT—GUESTS PUMMELING EACH OTHER
Into the melee walk the three stooges. They have canes on their arms and are carrying top hats.
CLOSER SHOT—THE STOOGES
They look at guests superciliously.
LONG SHOT—GUESTS SOCKING EACH OTHER
They are being knocked down, clothes are being torn off, etc.
CLOSE SHOT—THE STOOGES
MOE
(turning to boys)
My dear fellows—this is our punishment for associating with the hoi polloi.
LARRY
The self-same thoughts are mine concerning our unhappy situation.
(he turns to Curly)
What sayest thou?...
CURLY
(hastily)
The vicissitudes of one's circumstances in these surroundings makes it well nigh impossible to state one's feelings.
In the preceding as each stooge makes his initial speech . . . they put their top hats on jauntily, tapping them with one hand to set them at right angle.
CLOSE MOVING SHOT
as the Stooges start walking toward the exit. They are suddenly stopped by the two professors and another man.
RICH, LOVETT, & MAN
(chorus) (holding their fists in front of them)
Do you see that?
STOOGES
(chorus)
We do.
Professors and man haul off and do the roundhouse blow and drive top hats down over the Stooges' eyes. The boys start yelling as we
FADE OUT.[104]

FIGURE 36. The elite reveal an unexpected solidarity with the Stooges in *Hoi Polloi* (August 1935). "HOI POLLOI" © 1935, renewed 1963, Columbia Pictures Industries, Inc. All Rights Reserved. Courtesy of Columbia Pictures.

The comedic dynamics of class conflict are here dissolved into the unifying Depression-era theme of "the people." The intrusion plot serves no longer as a vehicle for comedic revenge on the world of privilege, but rather as a deconstruction of that privilege and proof of an underlying sameness (fig. 36). In their strained efforts to wear the mask of the elite, the Stooges unwittingly uncover a kind of knockabout solidarity that unites the classes: *All the World's a Stooge,* indeed—the title of one of the team's 1941 two-reelers.

But no single entry in the team's filmography better dramatizes these torsions and transformations than the aforementioned *Cash and Carry,* which explicitly yokes the intrusion plot to the theme of charity. The film's narrative effectively puts Stooge-style chaos in the service of charitable aid: the Stooges reduce a mansion to ruin as they search for buried treasure to finance a young boy's leg operation. Images of dependent children—crippled, starving, or both—had been central terms of charitable ideologies since the Progressive Era, and their resonance continued during the Depression. (Roosevelt himself helped maintain a rehabilitation institute for crippled children close to his presidential retreat in Warm Springs,

Georgia.) What distinguishes *Cash and Carry* in fact is the way the writers, Clyde Bruckman and Elwood Ullman, couch the narrative in loaded political symbols. The film begins, for instance, with a comedic representation of Depression-era impoverishment, as the Stooges return from a failed prospecting trip to their home in the city dump. In a draft of the first scene, Bruckman and Ullman explicitly locate the Stooges' residence in one of the "Hooverville" shantytowns built by homeless people during the period (so named after the president whose economic policies were blamed for the economic crisis) (fig. 37).

MED. SHOT—DUMP HEAP
There is a large pile of tin cans, alongside of which is constructed a make-shift tin 'HOOVERVILLE' shack. There is a stove pipe running at a crazy angle from the roof; smoke pouring from it. The Stooges drive into foreground, the car bucking; spits a couple of times and with a BANG comes to a stop.[105]

Entering the shack, the Stooges find it has been occupied by a crippled young boy and his sister, who are saving money for the brother's medical bills in a tin can. The script next includes a quip addressing the banking collapses of the early Depression years and the economy's recovery under Roosevelt:

MOE
You shouldn't leave your money lying around in a tin can. It oughta be in a bank!
LITTLE BOY
But will the bank give it back? . . .
CURLY
Oh, sure! They didn't used to—but they do now.[106]

This optimistic sense of a turnaround—"They didn't used to, but they do now"—is subsequently fulfilled in the film's conclusion, which offers an explicitly realized endorsement of the current president. In an example of somewhat creative geography, the Stooges accidentally tunnel from the mansion's basement into the United States Treasury, whereupon they are arrested and brought before FDR himself (an actor, back toward the camera). The president not only offers the Stooges clemency for the break-in but also agrees to arrange personally for the young boy's operation (fig. 38).

PRESIDENT'S RECEPTION OFFICE
There is a huge pair of doors, on front [sic] of which wait several dignified gentlemen in formal afternoon attire. A secretary comes through the doors.
SECRETARY
I'm sorry gentlemen, but the Senate Sub-committee will have to wait—the President is in conference.
He turns and opens the huge doors, starts to exit CAMERA TRUCKING [sic] WITH HIM THROUGH AND INTO the '*President's Office*,' disclosing the President seated at his desk—back to CAMERA. FACING CAMERA in front of the President

are the three Stooges, the girl and the crippled boy. (Note: We see only the President's back at all times.)

MOE

. . . and that's the way it happened Mr. President.

PRESIDENT

Oh, I see—

(turns to little boy)

Well, Jimmy, I shall arrange personally for your operation!

The group are elated, particularly Curly, who steps up to the desk and very intimately says to the President.

CURLY

Gee, Mr. President. You're sure a swell guy!

The President turns toward the Stooges.

PRESIDENT

. . . and as for you!

Curly jumps back in line as the Stooges stand, apprehensively.

PRESIDENT (continuing)

. . . in view of the extenuating circumstances I find it possible to extend you executive clemency.

CLOSEUP—CURLY

He reacts.

CURLY

Oh, Mr. President, please—not that!!

MED. SHOT—STOOGES

Moe jabs Curly in the stomach. Curly looks at Moe then back at the president.

CURLY

. . . but if that's your best offer, we'll take it.

We see the president chuckle as his shoulders heave in amusement. The Stooges elatedly wave goodbye, pick up the kid and start out of the office.

FADE OUT[107]

The initial specters of hardship and Hooverville are thus dissipated before the redeeming presence of Roosevelt himself, rendering the film's narrative a quasi-allegory of the nation's recent trajectory from collapse to recovery. Written in the spring of 1937, the film surely draws color from the short-lived optimism of those months, when Roosevelt's policies had helped bring industrial production back to pre-Depression levels. Roosevelt's function here is thus that of a great reconciler. His presidency not only affords resolution in narrative terms (he pardons the Stooges), but it also provides magical resolution to ideological fissures between philanthropy and federal aid as responses to economic crisis: aid comes from presidential decree, but a decree issued by the *president acting as individual philanthropist* ("I shall arrange personally").

In the process, however, *Cash and Carry* provides the Stooges' clearest demonstration of the limits of throwback slapstick as a vehicle for the ideological forms

FIGURES 37–38. From the hardships of "Hooverville" to the saving presence of FDR. Frame enlargements from *Cash and Carry* (September 1937).

of New Deal–era populism. The core of the film's comedy cleaves closely to the intrusion plot of so many of the team's films: the trio enters a prosperous domestic space and destroys it, here tearing out the walls and burrowing into the basement in their search for riches. Yet, atypically, the domestic space in question is empty and unoccupied: there is no society dinner in progress, no wealthy residents for the Stooges to exasperate. Instead, the mansion has long since been abandoned, sold to the team by a pair of con artists with a false promise of hidden treasure within. The destructive trope is thereby retained, but the destruction is no longer meaningfully directed against a social class; it is simply a deserted building that bears the brunt of the Stooge-induced chaos. It is as though the liberal populism enshrined in the New Deal–era theme of mutuality is unable to reconcile itself to the comic destructiveness typical of the Stooges, whose hostile energies must in consequence be displaced onto an empty room.

This, then, was the situation into which the Stooges and their writers found themselves forced at Columbia by their own comedic vested interests: to have committed to an "old-time" comedic mode that could engage the populist tropes of the New Deal only by withholding its underlying premises. Since the conception of class material was for the Stooges a conflictual one, their formulas could engage the forms of New Deal populism only by displacing the conflict that was their comedy's essential foundation—either by suggesting that class difference masks an unexpected sameness (the route of *Hoi Polloi* and others) or by literally hurling the Stooges' violence into a void (*Cash and Carry*). We have, in other words, the paradox of a raucous and divisive proletarianism struggling to reconcile itself with a vein of liberal populism that sought to suspend such divisiveness. Since the charitable motif was for the Stooges something like the negation of their comic dialectic, it would surface again in only a handful of further Stooges releases: in *Oily to Bed, Oily to Rise* (October 1939) and *Loco Boy Makes Good* (January 1942), both of which feature the team helping an old widow; in *Nutty but Nice* (June 1940), where the Stooges reunite a little girl with her kidnapped father; and finally in *Even as I.O.U.* (September 1942), in which the trio aids an evicted mother. But none of these films found any more satisfactory resolution to the contradictory populist logics of mutuality and class conflict, and as the New Deal order dissolved into the war years, so did these motifs disappear entirely from the Stooges films.[108] After a few remaining stabs at contemporaneity in the wartime satires *You Nazty Spy* (January 1940) and *I'll Never Heil Again* (July 1941), the Stooges settled back into the "old-time" destructiveness that was their stock-in-trade.

*

To bond together through shared laughter or to derive derisory satisfaction from comedy's aggressive instincts, to laugh with or to laugh at: these two possibilities circulate through the history and theory of comedy as opposed dynamics of comedic pleasure. There is, for instance, the theoretical dichotomy separating Soviet linguist Mikhail Bakhtin's festive paradigm of an inclusive "carnivalesque" laughter from more divisive models of laughter as a marker of boundaries, what Henri Bergson dubbed a "social gesture" to exclude the aberrant.[109] But there is also, we can now see, a corresponding *historical* alternation in the changing comedic modes of early twentieth-century American populism: on the one hand, the class-inclusive, sentimental Capra/Rogers/Laurel-Hardy/and so on paradigm that chimed with the political rhetoric of the New Deal; on the other, the class-conflictual, Sennett-Keystone template spawned from the more virulent class antagonisms of the turn of the century. It was the curious fate of the Three Stooges, I have been arguing, briefly to have tried to be both, just as it was the curious fate of Columbia's short-subjects division to have sustained the earlier style in a decade whose populist energies lay elsewhere. If Sennett-style slapstick was nostalgically recycled in the 1930s as merely "good old-fashioned" fun—as was the case at Columbia—this was because its moment as a vehicle of class symbolization no longer really held sway. This, then, is perhaps the final meaning of slapstick's nostalgic appropriation during this decade: that any style of comedy passes ineluctably toward the "old time" when the populism inherent in its form can no longer readily be made to agree with an existing political rhetoric.

We conclude with a remarkable text almost two decades later, in 1952, in which Jules White surveyed the changes to the short-subject market since his start at the studio. Written in the throes of Hollywood's postwar box-office slump—when profits declined from $120 million in 1946 to $30 million a decade later—the article evoked two-reel slapstick comedy as, by this point, both a virtually forgotten commodity and, at least according to White, a possible remedy for industry blues:

A visiting film exhibitor at Columbia Studios recently flabbergasted his guide by asking, "Have you a two-reel comedy shooting? I'd rather see one of those being made than anything else. On my theater marquee," added the showman, "I bill two-reelers right alongside a big AA feature. And do I pack them in!"

The sentiments of this exhibitor were right in line with those of Jules White, head of the shorts department at Columbia. He feels that these comedies are a vital and necessary part of the industry and that with the abolishment of double-bills, the two-reeler will return to the high position it held in the nostalgic past.

"Right now," says White, "during the wave of talk about the boxoffice decline, television, the worry of producing bigger and better pictures to meet the public's demands for strong entertainment or else, film folk can't seem to see the trees of two-reelers for the forest of 10-reelers. Their producers are relatively unknown, except to the oldtimers, and the fact that year after year, percentage-wise to their production cost, these comedies' profits compare favorably with the big budget hits, is a revelation."[110]

Again, the same constellation of meanings that this chapter has been tracing: the value of Columbia slapstick for outlying exhibition markets ("oldtimers") and the association of that value with nostalgia. The last of the major producers of short-subject slapstick, Jules White continued to bend his efforts to that market only through dreams of a return to the business conditions of the "nostalgic past," which, for him, referred to the era before the widespread diffusion of double features. But, by this late point, more had changed than business conditions alone; for if slapstick had long since passed into nostalgia, then one reason, this chapter has suggested, was that a certain connection between slapstick's raucous energies and their associated social meanings had long since become untied. Indeed, a sense of untimeliness was by this late point inscribed in Columbia's comedies as a formal element of film construction, as, faced with declining budgets, filmmakers now routinely cannibalized the unit's early films, not only for their plots but for entire scenes' worth of stock footage—a process that began in earnest with the Stooges' *The Pest Man Wins* (December 1951), which reused the pie-fight footage from *Half-Wits' Holiday* (January 1947), and which continued until the unit's termination in 1957. "Who the hell was going to care, anyway?" Jules White later reasoned, justifying the use of recycled footage. "Moe and Larry hadn't changed enough that anyone would notice. . . . The public never cared who did what in them."[111] Indeed, by the end, "new" Stooges films were typically assembled out of new scenes shot in just two or three days—sometimes just one—bridged with stock footage from earlier releases.[112] Curly had been replaced by Shemp, and Shemp, following his death in 1955, had in turn been replaced by Joe Besser; yet Larry and Moe continued, alternating from scene to scene in their final films with their younger selves. The production of short-subject slapstick at Columbia thus ended its days by collapsing explicitly into a self-perpetuating synchrony, a conflation of comedians past and comedians present that stood for the suspension of slapstick's relation to diachronic social time.

Coda

When Comedy Was King

If death is a condition of memorialization, then slapstick must have been dead by 1939. Such, at any rate, would be the lesson of Twentieth Century-Fox's Technicolor extravaganza *Hollywood Cavalcade,* a two-million-dollar nostalgia romp through the early days of the film industry, released in the fall of that year. "'Hollywood Cavalcade' is destined to arouse in the hearts and minds of millions of theatre-goers all the glamour and romance of the bygone era of motion pictures," the program for the film's October 4 premiere declaimed. "From the days of Keystone Cops . . . bathing beauties . . . slapstick comedies . . . to mighty modern motion picture masterpieces . . . here is a production which runs the gamut of Hollywood history!"[1] Hollywood had, of course, come very far from the early days of Keystone-style slapstick by the time Fox began production on this film (working title *Falling Stars*) the previous spring—so much so, in fact, that under Joseph Breen's watchful moral eye at the Production Code Administration, the filmmakers were not even allowed to use the word "pratfall," now deemed offensive.[2] Yet as the weighting of the program's rhetoric indicates, it was upon the recreation of Hollywood's supposedly slapstick-flavored youth that the film's nostalgic project was hung, and upon slapstick's subsequent displacement that its historiography was built. It is accordingly *Hollywood Cavalcade* that provides a first ending to my story.

*

The film was the second in a three-picture cycle of nostalgia musicals produced by Fox around this time—the others being *Alexander's Ragtime Band* (1938), a fictionalized account of the birth of ragtime, the success of which seems to have inspired the cycle, and *Swanee River* (1940), a biopic about minstrel song composer Stephen Foster. All three films were identical in terms of their narrative template, depicting the history of their respective popular cultural forms—ragtime, the movies, minstrel songs—as a socially integrative story of union and reconciliation, structured around a heterosexual romance.[3] It is thus as the story of one man's rise and fall and rise again, correlated with the vicissitudes of a love plot, that the narrative of *Hollywood Cavalcade* is arranged, offering a recounting of Hollywood's past that splices together elements of the lives of Mack Sennett, D. W. Griffith, and Cecil B. DeMille into the fictional career of one Michael Linnett Connors. (The protagonist's name is based on Sennett's birth name, Michael Sinnott). Journeying to Los Angeles in 1913, Connors (Don Ameche) rises quickly through the ranks of the early film industry, where, as an up-and-coming director, he accidentally "invents" the style of custard-pie slapstick and makes a comic star of his protégé, Molly Adair (Alice Faye), a Mabel Normand type who secretly loves him. Connors next pioneers the epic style of Griffithian historical spectacle after more serious ambitions take hold; in fact, it is while filming such a picture (with sets evocative of *Intolerance*) that Connors learns that Adair, tired of his workaholic tendencies, has married her costar, Nicky Hayden (Alan Curtis), sending Connors into a creative slump that derails his career. Resolution eventually comes years later, when Molly arranges Connors's return to the director's chair. Nicky unfortunately dies in an auto accident, leaving Molly and Connors free to declare their love for one another and boldly resolve to film a part-talkie together (it is now 1927). The film thus ends with a heterosexual reconciliation that restores Connors to his role as quasi-Promethean architect of early American film history: part Sennett, part Griffith, the inventor of film slapstick and the epic feature now also participates in the birth of the talkies. In the final shot, Molly and Connors embrace on a balcony overlooking Los Angeles, while, beside them, their producer intones: "It used to be a kind of game, the movies; and now look at it, a city . . . filled with people who make the entertainment for all the peoples of the world."

It is to this idea of early filmmaking as a "kind of game" that the early slapstick sequences are dedicated. Bearing the burden both of the film's verisimilitude as historical recreation and of its appeal as nostalgia, these scenes are populated with an assemblage of familiar silent comedians—Ben Turpin, Chester Conklin, and most centrally, Buster Keaton—all playing themselves. As a "Movie of the Week" report in *Life* magazine put it, producer Darryl F. Zanuck had "summoned out of the past" these "memorable personages" to "romp across the screen exactly as they used to do in the years of the movies' uproarious childhood."[4] (Mack Sennett also appears as himself somewhat later in the film, making a speech at Molly and Nicky's anniversary party—despite the fact that the fictional Michael Connors has appropriated much of his biography.) Nor was this the only gesture toward verisimilitude. According to studio publicity, the film's recreations of early slapstick, in these and other scenes, involved the appearance of an authentic period camera—"used by the Sennett Company twenty-five years ago" and dusted off for the occasion from "its seat of honor in the Los Angeles Museum's collection of movie relics"—and were staged with the technical assistance of erstwhile Sennett director Mal St. Clair, who instructed actors "to impersonate the famous comics of yesteryear."[5] Yet these appeals to authenticity here function as little more than props for the film's more aggressive mythologizing of slapstick's origins. We have already had cause to note—in the previous chapter—how the film participated in the broader misapprehension of Keaton as a veteran of slapstick's Keystone days.[6] More flagrant, though, is the film's depiction of slapstick's "birth." Keaton is called to set to perform a melodramatic scene with ingénue Adair and a mustachioed villain, whereupon he improvises an attack by picking up a custard pie left on the edge of the set and hurling it at his antagonist, accidentally hitting Molly instead. Keaton's improvised pie throwing not only disrupts the filming of the scene, brought to a halt so that Molly can clean up; it also breaks the scene's generic boundaries, pushing it outside of melodrama to provoke gales of laughter among the behind-the-camera personnel and inspiration for director Michael Connors. The film-within-the-film is subsequently completed not as melodrama but as a pie-throwing comedy. Next, in a two-minute sequence at a screening room, Connors's comedy is projected for his producers and, by proxy, us, the viewers: the film receives producer approval, and the studio places a daily order for five hundred custard pies at a local bakery. Thus, we are to believe, was slapstick invented. In the very process of "re-membering"—Barbara Myerhoff's term for the nostalgia that seeks the "reaggregation of . . . the figures who belong to one's life story" (here, early slapstick's)—the film instead significantly "*mis-remembers*" the past.[7]

Thus it was that the film's slapstick sequences were constructed—and, indeed, received—as the emblem of American cinema's youth, a cherished object of reminiscence abstracted from its own past. Without question, these sequences lay at the heart of *Hollywood Cavalcade*'s success—at least to judge from contemporary reviews—and they prompted a surprisingly unquestioning nostalgia from the majority of critics. A reviewer for the weekly *Variety* waxed lyrical, moved to reflect on slapstick as an object of loss: "Views of film making in the days when a pie was worth five jokes, and when Keystone cops whirled recklessly through traffic to save distressed maidens, are hilariously reproduced. Something more than pantomime passed from films when sound entered the studios. A complete form of storytelling labeled slapstick, also disappeared."[8] For the *Film Daily*'s critic, there was no contradiction between the film's nostalgic appeal and its authenticity as historical spectacle: "There is nostalgic footage a-plenty for woven into the romantic fabric is the 'birth' of the custard pie and Keystone Cop slapstick, the advent of the bathing beauty school of cinema, the 'discovery' of the De Millean type of spectacle and, finally, the debut of sound by way of Al Jolson and 'The Jazz Singer.'. . . . The parade of veterans—Mack Sennett, Lee Duncan [owner of Rin Tin Tin], Ben Turpin, Chester Conklin, among them—heightens the authenticity."[9] No less an authority on American cinema's past than Terry Ramsaye similarly celebrated the slapstick sequences for their authenticity, claiming, in fact, that the comedy *surpassed* that of the original object: "There is a magnificent sequence of the most utter of Keystone comedy slapstick," Ramsaye wrote in a special review in *Motion Picture Herald*. "The audience found it just as funny as the original audiences of old Keystone did—perhaps funnier. The pie throwing was equally authentic and successful."[10] Others, meanwhile, gave voice to a more interpreted nostalgia in explaining *Hollywood Cavalcade*'s appeal, linking the film's mood of reminiscence to the industry's present crises.[11] "Exhilarating, inspiring," one critic wrote. "[*Hollywood Cavalcade*] carries also a note of encouragement to picture production and exhibition personnel in these days when the industry faces, as it has faced before, a seemingly crucial period."[12]

So powerful was the film's recreation of Hollywood's early days as to prompt growing demand for a revival of old-style slapstick comedies and even rumors of Mack Sennett's return to filmmaking. "The great success of the early reels in 'Hollywood Cavalcade,'" commented the *Herald*, "is a demonstration that the public is in the mood for broad slapstick amusement."[13] Capitalizing on *Cavalcade*'s success, Sennett entered into discussions with Fox's production head, Darryl F. Zanuck, to produce a new Keystone-style feature, to be titled *Left at the Altar* or *Love in a Pullman Car*.[14] While nothing came of these plans, other companies were soon catering to the growing demand for throwback comedy. Shortly after *Cavalcade*'s release, RKO producer Max Gordon noted a "definite audience preference . . . for laughter" and described his intent to produce "laugh picture[s] . . . of the type that Lloyd and Chaplin used to make."[15] Meanwhile, a 1940 *Herald* article, titled "Trend toward Revival of Old Slapsticks," singled out two new firms specializing in slapstick reissues: the King of Comedy Film Corporation, which solely distributed Chaplin's Essanay films, and Motion Picture Jubilee Productions, offering comedies starring Ben Turpin, Snub Pollard, Mabel Normand, and others, "jazzed up with some 'screwball' commentary and sound effects." Commenting on the popularity of the films' screenings, Jubilee's president, Morton H. Miller, explained that older filmgoers were satisfying "a long sensed nostalgia for the old-time motion pictures, while the youthful film fan was relieved of a curiosity as to what silent pictures had to offer."[16] A belief in old-time slapstick's salvific potential had, by this point, become a frequently voiced element of mainstream movie culture—an ironic destiny for the subversive spirit with which Keystone-style slapstick had once confronted the very mainstream to which it now offered renewal.

*

But there it is again, in *Hollywood Cavalcade* too, the dividing line of sound—here the conclusion of a film whose operations as nostalgia depend upon a boundary beyond which Connor's and Adair's presumably continuing careers remain unrepresented and unrepresentable. Nostalgia, after all, requires a historiography grounded in rupture, in a change that permits neither of continuity (things were different then) nor of return (we can't go back); and in the historiography of slapstick, sound has long served that purpose, not only for the writers of *Hollywood Cavalcade* but for the broader legacy of twentieth-century scholarship on the form, extending from James Agee onward. As historian David Kalat astutely writes, "The history of the history of silent comedy has been a nostalgia industry since day one. By treating the advent of sound as a dividing line between Then and Now, any celebration of 'Then' turn[s] into a pining for something lost."[17] This book has been an alternative tale of that "something lost," focused instead on the social and industrial currents that transformed what had once been the very emblem of cinematic modernity into an anachronistic vehicle for old-time reminiscence. In the process of telling this tale, the significance of sound has changed, no longer as a technological division separating the "art" of pantomime from its sound-era freefall, but as a media change that introduced the principle of difference upon which silent-era nostalgia could thrive. The advent of sound may not have constituted a Rubicon moment in the history of cultural or industrial forms, but it did become the first example of a media change around which the *cultural memory* of succession and the resulting *affect of media nostalgia* would be organized, with slapstick enshrined as nostalgia's fast track.

As such, slapstick's fate bespeaks the degree to which Americans by the 1930s understood themselves to be living in the time of media. Cinematic forms like slapstick were now temporal markers for measuring cultural and, indeed, biographical life, "time machines," if you like, whose continued circulation either as pastiche or as reissues opened portals onto the youth of the movies as well as moviegoers' own youth. It is thus no accident that subsequent writers on the form so often engaged in a retrospective and personal recounting to clarify the stakes of their own investments.[18] James Agee's semiautobiographical novel, *A Death in the Family* (1957), returns to the slapstick tradition he had so famously analyzed in his 1949 essay, "Comedy's Greatest Era," only here in a spirit not of analysis but of remembrance: it opens with a boy Rufus and his father on a trip to the movies in 1915 to watch a Chaplin short, leaving the theater "wrapped in good humor, the memory of Charlie."[19] It is in the same spirit that Walter Kerr would commence his magisterial 1975 survey, *The Silent Clowns,* by proclaiming a primordial kinship with the great clowns of the 1920s. "When I first saw Keaton I didn't simply laugh at him, I fused with him, psyche locked to psyche; I recognized him as something known before birth, whatever that says about me—I have, as I say, yearned in what may be called a nostalgic way to see their films again."[20]

*

Yet an understanding of slapstick's fate as a nostalgia industry should reckon not only with the first-order nostalgia of those like Agee and Kerr who "lived through" the transition, but also with later audiences who, beginning in the 1950s, only ever knew slapstick second hand, as an already recycled form on television and other home-viewing technologies, and for whom it subsequently became part of *their* childhood. In the case of slapstick, the history of this later recycling properly deserves another book (from whose authorship I recuse myself), yet its outline further testifies to the role of media change in structuring nostalgia for media artifacts, here related not to technologically derived changes in representation (as in the case of sound) but to subsequent changes in slapstick's media accessibility (through television, home video, and the like). To live in the time of media, after all, does not refer solely to the way personal reminiscence may be periodized according to media change, but also to the process of indefinite recycling and remediation that makes media history a permanently renewable platform for the affective investments of later generations. The lineaments of that history provide the second ending to this book.

Television will be our opening port of call, with the Three Stooges the outstanding instance of how "old-time" slapstick came to be packaged as kiddies' TV in the late 1950s. Finally let go from Columbia's short subjects division in 1957—where they had starred in comic shorts for nearly a quarter century—the Stooges found unexpected new life on television, when, the following year, Columbia's Screen Gems division released the team's pre-1949 two-reelers for syndication. The sale attracted buyers in most of the nation's television markets, where the live-action shorts played on independent stations' weekday after-school schedules, creating what *Variety* described as a "highly favorable climate for slapstick" as children's programming.[21] Ratings alone testify to the team's startling resurgence as children's entertainers—in Chicago, for instance, where they achieved a 41 percent share for their time slot by December 1958—so much so that the trio began ending their live shows by thanking young audiences for bringing them "out of retirement."[22] Soon, television producers began capitalizing on the *Variety* reporter's intuitions by raiding public domain silent-era titles, comedies especially, to create re-edited clip compilations as family shows, such as Paul Killiam's ABC series *Silents, Please!* (1960–1961) and the syndicated children's programs *The Funny Manns* (1961) and *Fractured Flickers* (1963–1964), most of which featured comedy redubbing of silent films. Simultaneous with these was a spate of clip compilations, such as Harold Lloyd's own *Harold Lloyd's World of Comedy* (1962) and the stretch of eight compilation films produced by Robert Youngson: *The Golden Age of Comedy* (1958), *When Comedy Was King* (1960), *Days of Thrills and Laughter* (1961), *30 Years of Fun* (1963), *MGM's Big Parade of Comedy* (1964), *Laurel and Hardy's Laughing '20s* (1965), *The Further Perils of Laurel and Hardy* (1967), and *Four Clowns* (1970). Unexpectedly profitable, the films drew admiring commentary in the trade press for their success with a generationally diverse audience, "not only with the film buffs and oldsters, but with a whole new generation of film-conscious kids who will come to appreciate this comedy artistry for their very own. Who says there's no way to bridge the 'generation gap' today?"[23]

Yet the cycle of late 1950s/1960s-era slapstick reissues was not without consequence for the form's ongoing banalization. With the late 1950s revival of slapstick as after-school television, not only were the films now embedded more directly in children's culture than ever before but they were playing on a medium with a very different discursive identity from cinema, and this required negotiations regarding the limits of violent comedy as children's entertainment. As one reporter commented on the Stooges revivals in 1960, "Ever since the madcap Three Stooges zoomed . . . back into the limelight, . . . certain factions have been taking potshots at the trio's wholesale use of 'violence,'" adding that "the TV storm has been raised because television is an intimate medium that is going right into the home."[24] The recycling of slapstick as kiddies' programming thus stumbled over a discontinuity in media identities, forcing slapstick into the center of debates about child-appropriate limits to comic spectacle that resulted in one final, unexpected mutation in the form: its cartoonification. One strategy for children's television producers and programmers who sought to capitalize on the juvenile slapstick vogue and yet sidestep the critiques that live-action violence invited was to turn some of the best-known slapstick clowns into cartoon figures. Hanna-Barbera would be the most notable company to take up the reins here, producing *Laurel and Hardy* and *Abbott and Costello* cartoons in the mid-1960s. Meanwhile, Normandy Productions—headed by Norman Maurer, Moe Howard's son-in-law—launched *The New Three Stooges* animated series, which ran for over 150 episodes between 1965 and 1966, with live-action wraparounds featuring the aging trio (now Moe, Larry, and Curly-Joe De Rita). As Moe himself explained, the intent was to use animation as a way of redeeming knockabout as "safe" children's entertainment: the team, it was claimed, would tone down their eye gouging for the live-action introductions, instead consigning such "intentional" violence to the animated segments. One *Herald-Tribune* interview—titled "What Price Violence?"—clarified the strategy: "'When we're live,' Moe explained, "we'll cut out the . . . physical horseplay and substitute unintentional violence. . . . In other words, I won't purposely 'clunk' Curly or Larry, but if I'm carrying a ladder, let's say, and I make a quick turn, it could clip Curly on the bean—unintentionally. The deliberate stuff will be seen only in the cartoon segments.'"[25] It is as though animation, precisely because of its abstraction from material reality, could be considered an innocuous, indeed inoculating, format for knockabout violence.

And might one not hypothesize that these various renegotiations of slapstick's cultural meanings—peaking here at the very moment when baby-boom youngsters were inheriting the tradition—played a role in slapstick's ongoing consignment to the realm of presocial nostalgia? We here recall the process of "miniaturization" broached in our earlier discussion of *Babes in Toyland* (1934): miniature objects invite a reverie that "skews the experience of the social by literally *deferring* it" and so cannot be separated from an *a*social mode of apprehension.[26] And slapstick was nothing, by this point, if not fully "miniaturized": cut up into clips and accompanied by ironic and distancing commentary or cast into cartoon form wherein the clowns' infant-proportioned bodies now romped around a "plasmatic" fantasy land (fig. 39).[27] The form's thoroughgoing banalization (now by fixed association with kiddie audiences) was thus knotted on television screens to an ongoing miniaturization to produce a third term, *infantilization,* marking slapstick's consignment to a realm antecedent to the lived dialectics of social time.

The next port of call in our prospective history redeems this fate by virtue of another media switch—this time to the eight- and sixteen-millimeter home market for vintage films that, in the 1960s and 1970s, turned growing boomers into a new generation of collectors. Of the various companies trading in vintage film during this era (Classic Films, Ken Films, and so forth), Blackhawk commands attention as the company best positioned to exploit the slapstick market. Previously operating primarily as a rental library for films on sixteen millimeter with optical sound, Blackhawk altered its business model in 1952 by launching its own releases for sale under what it called its *Collector Series,* among which were Laurel and Hardy silents from the Hal Roach Studios and Keystone comedies licensed by Mack Sennett's one-time backer, Roy Aitken. By the time David Shepard joined the staff as vice president of product development in 1973, the company was nurturing collectors' enthusiasms with more than a dozen new releases every month, as well as establishing standards for restoration that were unprecedented in the vintage film market. Existing company correspondence from the 1970s—approximately a quarter of which deals with comedy—reveals Shepard providing information to inquiring collectors, mailing them copies of film historian Kalton C. Lahue's studies of silent slapstick, and consulting with Academy of Motion Picture Arts and Sciences archivist Sam Gill on the restoration of Chaplin's Mutual shorts (1916–1917) and Roscoe Arbuckle's *Fatty and Mabel Adrift* (January 1916).[28] We will thus also want to introduce to our prospective history figures like Lahue and Gill, whose publications—the coauthored history *Clown Princes and Court Jesters* (1970), as well as Lahue's lengthy streak of monographs on early cinema, including *Kops and Custards* (1968), *Mack Sennett's Keystone* (1971), and tellingly, *Collecting Classic Films* (1970)—became critical sources of information for the emerging collectors market, existing symbiotically with Blackhawk to restore a historicizing imperative to hobbyist enthusiasms.[29]

Perhaps a final reference to the work of literary theorist Susan Stewart can bring our argument home. In her groundbreaking study *On Longing,* Stewart introduces a celebrated distinction between the souvenir and the collection in terms of their respective relations to the past. The souvenir, she writes, "is not

3 Stooges Ready 'Fresh' TV Series

By SAMUEL D. BERNS

HOLLYWOOD, March 31. — The high ratings garnered by hundreds of old Three Stooges' comedies, now in their sixth run in many of the 156 markets showing the Screen Gems syndicated show, has inspired "Larry, Moe and Curly Joe" to produce their own "fresh" film series expressly for television titled "Three Stooges Scapbook."

In an interview with Moe Howard, who formed the Three Stooges' partnership more than 25 years ago, and Norman Maurer, producer of the new

series, it was pointed out that the new format will include a special added feature. In addition to the live action, which will occupy about 15 minutes of the half-hour program, two five-minute segments of animated cartoons are integrated in which the Stooges are cartoon characters carrying on their zany antics.

Howard called attention to the fact that the Stooges will eliminate any semblance of violence, such as "direct thumbing, pulling noses or slapping," which figured prominently in their old comedies.

"We are going directly into the home for family entertainment, and believe we can achieve high comedy without resorting to the kind of violence that kids are apt to imitate," Howard pointed out.

FIGURE 39. A trade report on Norman Maurer Productions' *Three Stooges Scrapbook*, one of the earliest efforts to repackage live-action slapstick in cartoon form for children's television. Samuel D. Berns, "3 Stooges Ready 'Fresh' TV Series," *Motion Picture Daily*, April 1, 1960.

simply an object appearing out of context, an object from the past incongruously surviving in the present"; rather, souvenirs are "magical objects" that "transform history into private time" by "envelop[ing] the present within the past" of one's own life experiences.[30] By contrast, the logic of the collection *extracts* from the past, both private and historical, in order to submit its artifacts to the present context of a classification: "The collection seeks a form of self-enclosure . . . [that] replaces history with *classification,* with order beyond the realm of temporality."[31] As souvenir, a past artifact has value as an anchor in one's own biographical encounters or experiences; as part of a collection, it serves as an item in a series in need of completion.

Scholars have qualified Stewart's distinction by suggesting that the two logics may in fact coexist (in the case, say, of a private collection), so that it would be more useful to think of the pairing as two poles on a spectrum rather than as exclusive alternatives.[32] Still, the applicability of the distinction to the case at hand seems clear. By the 1960s, the American slapstick tradition had long been a souvenir, arguably *the* preeminent souvenir of film history, its legacy sedimented for at least two generations as a portal into the private time of childhood and a token of a supposedly more authentic, primordial laughter ("when a pie was worth five jokes"). But the emergence of a collectors market spurred a shift in protocols toward classification and exhaustiveness, whose goal would be a perfect hermeticism. Undoubtedly such completion can never be materialized, given the archival absences that plague silent-era film, but it can be *symbolized,* its gaps "filled in" by a historical writing that stands in for the completion the collector desires. It is thus symptomatic, I think, that the preferred mode of scholarly expression for those boomer historians has come to be the filmography, a genre of slapstick historiography guided by rigorous protocols of identification that has reached a remarkable apex in recent years in publications like Brent Walker's magisterial resource *Mack Sennett's Fun Factory* (2010) and Richard Roberts's no less comprehensive cataloging of the Hal Roach Studios' output in *Smileage Guaranteed* (2013), as well as in Steve Massa's ongoing recovery of forgotten and marginalized comic performers, to name only some prominent examples.[33] Nostalgia transcends itself, it would seem, when the souvenir founds a collection, even when that collection takes virtual form only, as a catalogue raisonné.

This is why I rehearse this sketch for a prospective history, in order finally to locate *Hokum!* within the very cultural processes it has analyzed. Because I am somewhere in there too, a kid watching the PBS series *Harold Lloyd's World of Comedy* on BBC2 in the early 1980s—not a baby boomer (those are my parents) nor a collector, but one who has, like others, found in nostalgia the imperative to historicize, the longing mark that would overcome itself through historiography.

ABBREVIATIONS

AMPAS	Margaret Herrick Library, Academy of Motion Picture Arts and Sciences
BRTC	Billy Rose Theatre Collection, New York Public Library
EHMPW	*Exhibitors Herald and Moving Picture World*
EHW	*Exhibitors Herald-World*
EPS	Educational Press Sheets, Billy Rose Theatre Collection, New York Public Library
HRC	Hal Roach Collection, Performing Arts Archives, University of Southern California
JWC	Jules White Collection, Margaret Herrick Library, Academy of Motion Picture Arts and Sciences
MPH	*Motion Picture Herald*
MPN	*Motion Picture News*
MSC	Mack Sennett Collection, Margaret Herrick Library, Academy of Motion Picture Arts and Sciences
NYHT	*New York Herald-Tribune*
NYT	*New York Times*
RBS	Robert Benchley Scrapbooks, Billy Rose Theatre Collection, New York Public Library
RKO	RKO Studio Collection, Special Collections, University of California, Los Angeles
USC	Performing Arts Archives, University of Southern California
WB	Warner Bros. Archives, Performing Arts Archives, University of Southern California

NOTES

INTRODUCTION

1. Percy W. White, "Stage Terms," *American Speech* 1, no. 8 (May 1926): 436.

2. Edna Ferber, *Cimarron* (New York: Doubleday, Doran, 1929), 207.

3. Frank J. Wilstach, "How 'Hokum' Declined from Its High Estate," *NYT*, July 8, 1928, 96.

4. "A Stage Dictionary," *NYT*, September 9, 1923, X2.

5. White, "Stage Terms," 436.

6. "The Appeal of the Primitive Jazz," *Literary Digest,* August 25, 1917, 28.

7. "Al Jolson's 'Hokum Stuff,'" *Philadelphia Evening Public Ledger,* February 13, 1915, 5. I am grateful to Paul Babiak for uncovering this reference.

8. Walter Winchell, "A Primer of Broadway Slang," *Vanity Fair,* November 1927, 132.

9. Wilstach, "How 'Hokum' Declined," 96.

10. See https://books.google.com/ngrams (accessed July 25, 2016). Ngram is an application that charts word frequency across texts scanned into Google Books.

11. David Kalat, "Don't Send in the Clowns," *Movie Morlocks,* September 21, 2013 http://moviemorlocks.com/2013/09/21/dont-send-in-the-clowns (accessed July 25, 2016).

12. James Agee, "Comedy's Greatest Era," *Life,* September 5, 1949, 70.

13. Ibid., 71.

14. Gerald Mast, *The Comic Mind: Comedy and the Movies* (Indianapolis, IN: Bobbs-Merrill, 1973), 199.

15. Ibid.

16. Ibid., 202.

17. Walter Kerr, *The Silent Clowns* (New York: DaCapo Press, 1975), 3, 25.

18. Rudolph Arnheim, *Film as Art* (Berkeley: University of California Press, 1957 [1933]), 109, 111.

19. Kerr, *The Silent Clowns,* 48.

20. Ibid., 337–38.

21. The term is Raymond Williams's, from his critique of technological determinism in *Television: Technology and Cultural Form* (London: Routledge, 2003 [1974]), 133.

22. Ian Hutchby, "Technologies, Texts, and Affordances," *Sociology* 35, no. 2 (2001): 441–56.

23. Capra's account is handily debunked by Joseph McBride's *Frank Capra: The Catastrophe of Success* (New York: St. Martin's Press, 1992), 139–79.

24. Pierre Bourdieu, *The Rules of Art: Genesis and Structure of the Literary Field*, trans. Susan Emanuel (Stanford, CA: Stanford University Press, 1996 [1992]), 181.

25. The term "virtual Broadway" is from Donald Crafton, *The Talkies: American Cinema's Transition to Sound, 1926–1931* (Berkeley: University of California Press, 1999), ch. 3.

26. White quoted in Ted Okuda and Edward Watz, *The Columbia Comedy Shorts: Two-Reel Hollywood Film Comedies, 1933–1958* (Jefferson, NC: McFarland, 1986), 19.

27. Such an approach is the basic tenet of what sociology calls "Actor-Network-Theory," to which an invaluable introduction can be found in Bruno Latour, *Reassembling the Social: An Introduction to Actor-Network-Theory* (Oxford: Oxford University Press, 2005).

28. "What the Picture Did for Me," *MPH*, July 13, 1935, 85; "What the Picture Did for Me," *EHW*, March 9, 1929, 71; "What the Picture Did for Me," *MPH*, March 31, 1934, 67.

29. Charles Wolfe, "On the Track of the Vitaphone Short," in Mary Lea Bandy, ed., *The Dawn of Sound* (New York: Museum of Modern Art, 1989), 38.

30. For the relevant portion of the Motion Picture Code, see "Complete Text of the Code," *MPH*, December 2, 1933, 32. Although the NIRA codes were declared unconstitutional two years later, the major studios continued to permit varying degrees of exhibitor selection in the booking of short subjects, and none made shorts contingent on feature rentals. A 1938 *Motion Picture Herald* article collates the major distributors' approaches to short-subject bookings, ranging from the liberal (Twentieth Century–Fox: "If a particular operating policy calls for no shorts, then we do not insist on shorts"; MGM: "Where a theatre [does] not use shorts the company [does] not demand their purchase") to the harder sell (RKO: "Accounts are persuaded to take as many as they can absorb"; Universal: "Our salesmen naturally try to sell as much of our short subject product as possible"). "Federal Invitation Answered; Steffes Sees New Divorce Bill," *MPH*, October 1, 1938, 17–18.

31. Besa Short, "Short Circuit," *MGM Shortstory*, October 1939, 6. The Interstate Circuit's innovations were well covered in the trade press: "They Screen Every Short," *MPH*, May 16, 1936, 86; "McFadden Heads New Shorts Unit," *MPH*, April 10, 1937, 44; "Interstate Claims Solution for Duals by 'Balancing' of Shorts," *MPH*, July 17, 1937, 36; Fred McFadden, "Short Success Story," *MGM Shortstory*, January–February 1938, 4–6; "All-Shorts Program Triples Dallas Theatre Business," *MPH*, August 26, 1939, 69.

32. Jennifer Fleeger, *Sounding American: Hollywood, Opera, and Jazz* (New York: Oxford University Press, 2014); Katherine Spring, *Saying It with Songs: Popular Music and the Coming of Sound to Hollywood Cinema* (New York: Oxford University Press, 2013).

33. The one exception here is Peter Brunette's Derridean analysis "The Three Stooges and the (Anti-)Narrative of Violence: De(con)structive Comedy," in Andrew S. Horton, ed., *Comedy/Cinema/Theory* (Berkeley: University of California Press, 1991), 174–87.

34. Raymond Williams, *Marxism and Literature* (New York: Oxford University Press, 1977), 122–23.

35. Ibid., 124.

36. Paula S. Fass, *The Damned and the Beautiful: American Youth in the 1920's* (New York: Oxford University Press, 1977). See also Judith Yaross Lee, *Defining New Yorker Humor* (Jackson: University of Mississippi Press, 2000), which engages Fass's ideas in relation to the early *New Yorker* humorists.

37. Bourdieu, *The Rules of Art*, 214.

38. "Sound Places Shorts on Feature Plane, Says Al Christie," *EHW*, June 15, 1929, 128.

39. The concept of banalization is addressed in Bourdieu's *The Rules of Art*, 205–6.

40. Max Weber, *Economy and Society: An Outline of Interpretive Sociology*, vol. 2, trans. Ephraim Fischoff (New York: Bedminster Press, 1968 [1924]), 937, referred to in my book *The Fun Factory: The Keystone Film Company and the Emergence of Mass Culture* (Berkeley: University of California Press, 2009), 12.

41. I am indebted to Frank Kelleter for insisting on this point to me.

42. On Bourdieu's concept of "social aging," see ch. 3.

43. See, for example, Michele Hilmes, *Radio Voices: American Broadcasting, 1922–1934* (Minneapolis: University of Minnesota Press, 1997); Ross Melnick, *American Showman: Samuel "Roxy" Rothafel and the Birth of the Entertainment Industry* (New York: Columbia University Press, 2012); Spring, *Saying It with Songs*; Steve J. Wurtzler, *Electric Sounds: Technological Change and the Rise of Corporate Mass Media* (New York: Columbia University Press, 2007).

44. Susan Stewart, *On Longing: Narratives of the Miniature, the Gigantic, the Souvenir, the Collection* (Durham, NC: Duke University Press, 1993), 65.

45. Barbara Myerhoff, "Life History among the Elderly: Performance, Visibility, and Re-membering," in M. Kaminsky, ed., *Remembered Lives: The Work of Ritual, Storytelling, and Growing Older* (Ann Arbor: University of Michigan Press, 1992), 240.

46. Mack Sennett anticipated the anthology trend by a decade with his own slapstick compilation film, *Down Memory Lane*, released by Eagle-Lion in the summer of 1949.

47. See Henry Jenkins and Kristine Brunovska Karnick, "Introduction: Golden Eras and Blind Spots—Genre, History and Comedy," in Jenkins and Karnick, eds., *Classical Hollywood Comedy* (New York: Routledge, 1995), 1–2; and David Kalat, "The History of the History of Silent Comedy," *Movie Morlocks*, February 18, 2012 http://moviemorlocks.com/2012/02/18/the-history-of-the-history-of-silent-comedy (accessed July 25, 2016) for an analysis of this trend.

48. Jenkins and Karnick, "Introduction: Golden Eras and Blind Spots," 2.

CHAPTER 1. "THE CUCKOO SCHOOL"

1. "News of Theatres," *New York Tribune*, April 3, 1923, 8.

2. "Lewis and Dody Capture Honors at Loew's Grand," *Atlanta Constitution*, January 12, 1926, 8.

3. Vivian Shaw, "The Cuckoo School of Humour in America," *Vanity Fair*, May 1924, 46. The preceding description of Lewis and Dody's performance is based partly on this essay (discussed further below), as well as on the team's 1922 phonograph recording of *Hello! Hello! Hello!* (Columbia A3783). Thanks to Michael Cumella for making this recording available to me.

4. My use of speculative conjecture aligns these opening paragraphs with the practices of literary new historicism, whose productivity for the study of early twentieth-century popular culture has recently been explored by Donald Crafton. See his essay, "McKay and Keaton: Colligating, Conjecturing, and Conjuring," *Film History* 25, nos. 1–2 (2013): 31–44, as well as his forthcoming work on Joseph P. Kennedy. On the new historicism, see Catherine Gallagher and Stephen Greenblatt, *Practicing New Historicism* (Chicago: University of Chicago Press, 2000).

5. Gilbert Seldes, *The 7 Lively Arts,* 2nd ed. (New York: Sagamore Press, 1957 [1924]), 294–95, 207.

6. Ibid., 295.

7. Quoted in Ann Douglas, *Terrible Honesty: Mongrel Manhattan in the 1920s* (New York: Farrar, Straus and Giroux, 1995), 69.

8. Seldes, *The 7 Lively Arts,* 31, 16.

9. Ibid., 32.

10. Ibid., 20.

11. Ibid., 142, 179, 221–22.

12. Shaw, "The Cuckoo School of Humour in America," 46, 98.

13. Ibid., 46.

14. Ibid. One of the teams influenced by Lewis and Dody may well have been Al Shaw and Sam Lee, whose performances in early sound shorts—first at Vitaphone, in *The Beau Brummels* (ca. September 1928) and *Going Places* (June 1930), subsequently in a one-off for MGM, *Gentlemen of Polish* (June 1934)—could similarly be described as a "combination of apparent dullness with insanity."

15. Ibid.

16. Ibid.

17. Seldes, *The 7 Lively Arts,* 14.

18. Alan Devoe, "The Passing of the Joke," *Vanity Fair,* June 1934, 16d. The list of terms is taken from Max Eastman, *Enjoyment of Laughter* (New York: Simon and Schuster, 1936), 177.

19. George Jean Nathan, *The Popular Theatre* (New York: Alfred A. Knopf, 1918), 129; George Jean Nathan, "The Chaplin Buncombe," in *Passing Judgments* (New York: Alfred A. Knopf, 1935), 212.

20. The quoted characterizations are from, respectively, Gian Piero Brunetti's remarks at the Museum of Modern Art's 1985 "Slapstick Symposium," quoted in Eileen Bowser, ed., *The Slapstick Symposium* (Brussels: Fédération Internationale des Archives du Film, 1988), 90; and Miriam Hansen, "The Mass Production of the Senses: Classical Cinema as Vernacular Modernism," *Modernism/Modernity* 6, no. 2 (1999): 70.

21. On early mischief gag films, see Tom Gunning, "Crazy Machines in the Garden of Forking Paths: Mischief Gags and the Origins of American Film Comedy," in Kristine Brunovska Karnick and Henry Jenkins, eds., *Classical Hollywood Comedy* (New York: Routledge, 1995), 87–105. On Keystone's "uproarious inventions," see my essay "'Uproarious Inventions': The Keystone Film Company, Modernity, and the Art of the Motor," in Tom Paulus and Rob King, eds., *Slapstick Comedy* (New York: Routledge, 2010), 114–36.

22. See Miram Bratu Hansen, *Cinema and Experience: Siegfried Kracauer, Walter Benjamin, and Theodor W. Adorno* (Berkeley: University of California Press, 2012), esp. ch. 2.

23. Siegfried Kracauer, "Artistisches und Amerikanisches," *Frankfurter Zeitung,* January 29, 1926, quoted in Hansen, *Cinema and Experience,* 47.

24. Slapstick's claim to modernity was qualified by the mid-1920s in another way, too, albeit one that received no commentary from observers of the time; namely, slapstick had long ceased to be a welcoming venue for female clowns who challenged normative constructions of gendered deportment. As film historian Steve Massa explains, "The heyday of the female slapstick clowns had been in the Teens, when Fay [Tincher], Alice Howell, Louise Fazenda, Gale Henry, Polly Moran, etc., all emerged and had their own starring series. In the early 1920s, the mode had changed to leading ladies (Dorothy Devore, Alice Day, Wanda Wiley) who could perform physical comedy and all the more eccentric women had to find new venues—Louise Fazenda, Gale Henry, and Polly Moran migrated to character roles in features, Alice Howell retired and Fay [Ticher] found refuge at Universal." Steve Massa, *Lame Brains and Lunatics: The Good, the Bad, and the Forgotten of Silent Comedy* (Albany, GA: BearManor Media, 2013), 169. During the first decade of sound shorts, the number of female-led comedies dwindled further still to a single season of Louise Fazenda comedies at RKO (1930–1931), a series starring vaudevillian Lulu McConnell at Paramount (1931–1932), a "mini-series" of two Polly Moran shorts at Columbia (1936–1937), and, most notably, five seasons of Hal Roach shorts pairing Thelma Todd first with Zasu Pitts (1931–1933), then with Patsy Kelly (1933–1936). An insightful analysis of silent-era female clowns and changing ideologies of femininity is provided by Kristen Anderson Wagner, "Pie Queens and Virtuous Vamps: The Funny Women of the Silent Screen," in Andrew Horton and Joanna Rapf, eds., *A Companion to Film Comedy* (Malden, MA: John Wiley, 2013), 39–60.

25. Jacky Bratton and Ann Featherstone, *The Victorian Clown* (Cambridge: Cambridge University Press, 2006), 4.

26. Walter Blair and Hamlin Hill, *America's Humor: From Poor Richard to Doonesbury* (New York: Oxford University Press, 1978), 368.

27. Carl van Doren, "Day In and Day Out: Manhattan Wits," in van Doren, ed., *Many Minds* (New York: Albert A. Knopf, 1924), 183.

28. Paula S. Fass, *The Damned and the Beautiful: American Youth in the 1920's* (New York: Oxford University Press, 1977). The applicability of Fass's ideas to an understanding of the *New Yorker* humorists is suggested by Judith Yaross Lee, *Defining New Yorker Humor* (Jackson: University of Mississippi Press, 2000), 24–25.

29. Fass, *The Damned and the Beautiful*, 124, 122.

30. Lee, *Defining New Yorker Humor*, 24.

31. Quoted in ibid., 34.

32. Robert Benchley, "The Brow-Elevation in Humor," in *Love Conquers All* (New York: Henry Holt, 1922), 304, 306.

33. Quoted in Douglas, *Terrible Honesty*, 39.

34. The notion of "position taking" I derive from Pierre Bourdieu, who uses it to describe the process whereby new literary movements demarcate themselves through a stance of rejection and exclusion with regard to previous traditions and their publics: "When a new literary or artistic group imposes itself on the field, the whole space of positions and the space of corresponding possibilities . . . find themselves transformed because of it: with its accession to existence, that is, to difference, the universe of possible options finds itself modified, with formerly dominant productions, for example, being downgraded to the status of an outmoded or classical product." Pierre Bourdieu, *The Rules of Art: Genesis and Structure of the Literary Field*, trans. Susan Emanuel (Stanford, CA: Stanford University Press, 1995), 234.

35. Parker quoted in Douglas, *Terrible Honesty*, 37.

36. A case in point would be E. B. White's celebrated 1925 essay, "Child's Play," in which the author recounted an experience at a crowded restaurant when a waitress spilled buttermilk down his neck. E. B. White, "Child's Play," *New Yorker*, December 26, 1925, 17. On "little man" humor, see Stephen H. Gale, "Thurber and the *New Yorker*," *Studies in American Humor* 3, no. 1 (Spring 1984): 11–23; Lee, *Defining New Yorker Humor*, passim.

37. See Lee, *Defining New Yorker Humor*, 43, for a brief discussion.

38. Paolo Virno, *Multitude: Between Innovation and Negation*, trans. Isabella Bertoletti, James Cascaito, and Andrea Casson (Los Angeles: Semiotext(e), 2008), 146–48.

39. Robert Benchley, "All About the Silesian Problem," in *Love Conquers All*, 117; Donald Ogden Stewart, "How We Introduced the Budget System into Our Home," *New Yorker*, June 2, 1928, 24; Baer quoted in Seldes, *7 Lively Arts*, 249.

40. Frank Sullivan, "Should Admirals Shave?" *New Yorker*, February 25, 1928, 17.

41. Robert E. Rogers, Lecture Outline VIII, "Literature for Recreation, Part II: The Literature of Wit and Humor," Division of University Extension, Massachusetts Department of Education (1932–1933), RBS.

42. Stephen Leacock, *Humor: Its Theory and Technique* (Binghamton, NY: Dodd, Mead, 1935), 94.

43. See Robert Bernard Martin, *The Triumph of Wit: A Study of Victorian Comic Theory* (Oxford: Clarendon Press, 1974). As Leacock put it, parroting those Victorian theorists, humor "in its highest form" should bypass "mere language" in order to represent "an outlook upon life, . . . in which the fever and fret of our earthly lot is contrasted with its shortcomings." Leacock, *Humor*, 261. On Leacock's views on humor, see Gerald Lynch, *Stephen Leacock: Humour and Humanity* (Kingston, Can.: McGill-Queen's UP, 1988).

44. Eastman, *Enjoyment of Laughter*, 254.

45. Ibid., 3. If this is ultimately a rather quibbling distinction, it is because Eastman's model is otherwise identical to Freud's. Barring the issue of whether nonsense "counts" as humor (Freud, no; Eastman, yes), both theorists comprehend enjoyment of humor as fundamentally atavistic.

46. See n. 82 on the Hawaiians routine.

47. Eastman, *Enjoyment of Laughter*, 256.

48. Sanford Pinsker, "On or about December 1910: When Human Character—and American Humor—Changed," in William Bedford Clark and W. Craig Turner, eds., *Critical Essays on American Humor* (Boston: G. K. Hall, 1984), 184–99.

49. Leonard Diepeveen, ed., *Mock Modernism: An Anthology of Parodies, Travesties, Frauds, 1910–1935* (Toronto: University of Toronto Press, 2014).

50. Quoted in ibid., 218.

51. Examples would include Franklin P. Adams, "The Conning Tower," *Chicago Evening Post*, April 9, 1913, 8; and "To the Neo-Pseudoists," in *By and Large* (Garden City, NY: Doubleday, Page, 1914), 84; James Thurber, "More Authors Cover the Snyder Trial," *New Yorker*, May 7, 1927, 69; and E. B. White, "Is a Train," *New Yorker*, October 27, 1934, 26—all found in Diepeveen, *Mock Modernism*.

52. Max Eastman, *The Literary Mind: Its Place in an Age of Science* (New York: Charles Scriber's Sons, 1931), 63–64. The "Cult of Unintelligibility" essay appears in *Harper's Magazine*, April 1, 1929, 632–39.

53. Eastman, *The Literary Mind*, 76.

54. Douglas, *Terrible Honesty*, 21.

55. Put another way, cuckoo humor cannot be read as a comedic practice that simply "reproduced" class distinction, as a vulgar Bourdieusianism would have it, but as a linguistic form that partook in the broader restructuring of metropolitan entertainments and social life across lines of ethnicity, race, and sexuality. On the pansy craze, see George Chauncey, *Gay New York: Gender, Urban Culture, and the Making of the Gay Male World, 1890–1940* (New York: Basic Books, 1994), ch. 11; and Chad Heap, *Slumming: Sexual and Racial Encounters in American Nightlife, 1885–1940* (Chicago: University of Chicago Press, 2009), ch. 6.

56. The process exemplifies what Pierre Bourdieu has dubbed the "social ranking" of geographical space. "[A] group's real social distance from certain assets must integrate the geographical distance, which itself depends on the group's spatial distribution and, more precisely, its distribution with respect to the 'focal point' of economic and cultural values. . . . Thus, the distance of farm workers from legitimate culture would not be so vast if the specifically cultural distance implied by their low cultural capital were not compounded by their spatial dispersion. Similarly, many of the differences observed in the (cultural and other) practices of the different fractions of the dominant class are no doubt attributable to the size of the town they live in." Pierre Bourdieu, *Distinction: A Social Critique of the Judgment of Taste*, trans. Richard Nice (Cambridge, MA: Harvard University Press, 1984), 124.

57. This paragraph adapts points from David Kalat, "The 1927 Effect," *Movie Morlocks*, August 11, 2012 http://moviemorlocks.com/2012/08/11/the-1927-effect (accessed May 28, 2015).

58. Mordaunt Hall, "The Screen," *NYT*, February 8, 1927, 21. Tom Dardis claims, "The domestic gross was only $474,264, over $300,000 less than his previous film, *Battling Butler*." Dardis, *Keaton, the Man Who Wouldn't Lie Down* (New York: Scribner, 1979), 145. However, Dardis is comparing apples to oranges here, since the cited *Battling Butler* gross is worldwide.

59. *New York World* quoted in Joseph McBride, *Frank Capra: The Catastrophe of Success* (New York: St. Martin's Press, 1992), 178.

60. He did not revive the film until 1969 and mentioned it but once in his autobiography. Box-office figures taken from Charles Maland, *Chaplin and American Culture: The Evolution of a Star Image* (Princeton, NJ: Princeton University Press, 1989), 110.

61. Advertisement for *The Circus, MPN*, January 7, 1927, 19.

62. Theodor Huff, *Charlie Chaplin* (New York: Henry Schuman, 1951), 210.

63. On metropolitan critics' revolt against sentimentalism, see Lea Jacobs, *The Decline of Sentiment: American Film in the 1920s* (Berkeley: University of California Press, 2008). Chaplin's embrace by genteel critics has been widely observed but is best contextualized in Maland, *Chaplin and American Culture*, chs. 1 and 2.

64. Heywood Broun, "It Seems to Me," *New York Telegram*, November 27, 1928, n.p., *Animal Crackers* clippings file, BRTC.

65. Virginia Morris, "We—Want—Hokum!" *Picture Play*, April 1927, 20.

66. Ibid., 20–21.

67. George Jean Nathan, "The National Humor," in *Comedians All* (New York: Alfred A. Knopf, 1919), 237–49.

68. Rowland Barber, *The Night They Raided Minsky's: A Fanciful Expedition to the Lost Atlantis of Show Business* (New York: Simon & Schuster, 1960), 205. The material in this paragraph is condensed from Andrew Davis, *Baggy Pants Comedy: Burlesque and the Oral Tradition* (New York: Palgrave Macmillan, 2011), ch. 3.

69. Gilbert Seldes, "Fat Ladies," *New Republic*, March 30, 1932, 183.

70. The term "innovative nostalgia" is from Robert M. Crunden, *Ministers of Reform: The Progressives' Achievement in American Civilization, 1889–1920* (Urbana: University of Illinois Press, 1984).

71. This palimpsest quality was neatly foregrounded in a group routine from the show in which, looking for acting jobs, they all in turn gave imitations of Gallagher and Shean. The brothers would adapt the bit for the film *Monkey Business* (1931), in a series of imitations of Maurice Chevalier. (There is, incidentally, a family in-joke in the version of this sketch from *I'll Say She Is*: Al Shean was the Marx Brothers' uncle.)

72. Percy Hammond, "The Theatres," *New York Herald*, May 20, 1924, n.p., *I'll Say She Is!* clippings file, BRTC.

73. Groucho Marx, *The Groucho Phile: An Illustrated Life* (New York: Gallahad Books, 1976), 50, quoted in Frank Krutnik, "Mutinies Wednesdays and Saturdays: Carnivalesque Comedy and the Marx Brothers," in Horton and Rapf, eds., *A Companion to Film Comedy*, 98–99. On the vaudeville team Bickel, Watson, and Wrothe, see Steve Massa, *The Mishaps of Musty Suffer: DVD Companion Guide* (New York: Undercrank Productions, 2014). Following an acclaimed career in vaudeville and on Broadway, Harry Watson eventually appeared in film in three series of "Musty Suffer" one-reel comedies released by George Kleine from 1916 to 1917.

74. Alexander Woollcott, "Harpo Marx and Some Brothers," *New York Sun*, May 20, 1924, n.p., *I'll Say She Is!* clippings file, BRTC.

75. Krutnik, "Mutinies Wednesdays and Saturdays," 97. I am indebted to Krutnik's superb analysis for some of the interpretive framework of this and the previous paragraph.

76. Already by the time of the Broadway production of *Animal Crackers*, these creative alliances were drawing amused commentary among the critics' peers. Writing in the *New York Review*, Colgate Baker wondered aloud whether the Marxes' critical adulation betrayed a suspicious conflict of interests. "The most enthusiastic Marxian fans and shouters are the drama critics (with one exception), the drama editors and the columnists (without any exception). The fact that most of the critics, editors and columnists have had a hand in writing some of the show, of course has nothing to do with their feelings. The spirit of altruism that prevails these days, the detached, calm, unbiased poise of our criticism is too well known that maybe the boys are unconsciously influenced in their motivation—perish the thought!" Colgate Baker, "The Marvelous Marx Brothers and 'Animal Crackers,'" *New York Review*, Dec. 22, 1928, n.p., *Animal Crackers* clippings file, BRTC. The "one exception" mentioned by Baker would seem to be the *New York World* critic St. John Ervine, who, among Manhattan's critics, published the lone condemnatory review, in "The New Play," *New York World*, October 25, 1928, n.p., *Animal Crackers* clippings file, BRTC.

77. Krutnik, "Mutinies Wednesdays and Saturdays," 99.

78. Kaufman quoted in Stefan Kanfer, *Groucho: The Life and Times of Julius Henry Marx* (New York: Alfred A. Knopf, 2000), 93.

79. Perelman quoted in Sanford Pinsker, "Perelman: A Portrait of the Artist as an Aging *New Yorker* Humorist," *Studies in American Humor* 3, no. 1 (Spring 1984): 48.

80. John Anderson, "'Animal Crackers,' A New Show, Opens at 44th St.," *New York Evening Journal*, n.d., n.p., *Animal Crackers* clippings file, BRTC.

81. Robert Benchley, "The Theatre," *Life*, November 17, 1928, n.p.; and Gilbert Seldes, "Markert and the Marxes," *New Republic*, November 14, 1928, n.p., *Animal Crackers* clippings file, BRTC.

82. "The Theatre: Loudest and Funniest," *Time*, October 24, 1920, quoted in Krutnik, "Mutinies Wednesdays and Saturdays," 99.

83. "Joe Cook a 'Whole Show,'" n.s., April 11, 1916, n.p., Joe Cook clippings file, BRTC.

84. An article on Cook's 1942 retirement recalls the genesis of this famous bit. "His best known gag was his debate with himself as to whether he would imitate four Hawaiians. He recalled once that he had conceived it while playing in Akron, Ohio. He started it by saying, 'I will now imitate two Hawaiians' and then 'monkeying around' while playing a ukulele. Then he said: 'I could imitate four Hawaiians, but I won't.' Then, realizing he would have to tell the audience, now warmed up, why he wouldn't, he said: 'Ladies and gentlemen, you have just seen me imitate two Hawaiians, so why should I do four Hawaiians and show up those who can do only two Hawaiians and possibly be the cause of losing their positions? It's a principle with me.' When even the orchestra laughed, Mr. Cook realized the gag was good. He kept it and elaborated on it." "Joe Cook Quits Stage Because of Poor Health," *NYHT*, February 5, 1942, n.p., Joe Cook clippings file, BRTC.

85. Edgar Selwyn, "Speaking of Talking Pictures," *Theatre Magazine*, January 1930, 30, quoted in Henry Jenkins, *What Made Pistachio Nuts? Early Sound Comedy and the Vaudeville Aesthetic* (New York: Columbia University Press, 1992), 94.

86. Jenkins, *What Made Pistachio Nuts?*, 94–95.

87. Ibid., 278.

88. Clark recalled his experiences with Hollywood writers in misogynistic terms: "Clark insisted that he sit in on the writers' conferences, and his experiences there left a mark on him. He would look polite and interested while the former conductor of a bird column for an Arkansas weekly and a female author of a libidinous best-seller outlined a number of hilarious suggestions, then quietly reply, 'No.'" "Profiles, III—Up from Moose Jaw," *New Yorker*, September 27, 1947, 40.

89. "Profiles, II—Minstrel and Circus Days," *New Yorker*, September 20, 1947, 34.

90. "Profiles, III," 36. "Terrier" is from "Bobby Clark's Humor in a Class by Itself," *New York World-Telegram*, September 12, 1942, n.p., Bobby Clark clippings file, BRTC.

91. "Profiles—Comedian," *New Yorker*, September 13, 1947, 36.

92. "Architects of Laughter," *New York World-Telegram*, March 25, 1932, n.p., Bobby Clark clippings file, BRTC.

93. J. Brooks Atkinson, "Various Current Matters," *NYT*, October 1, 1926, X1; Bide Dunley, "The Ramblers," *New York Evening World*, September 21, 1926, n.p., *The Ramblers* clippings file, BRTC; "Bobby Clark Comes Rambling," *New York Sun*, September 21, 1926, n.p., *The Ramblers* clippings file, BRTC. *The Ramblers* was said at the time to be the first show of its size to have paid off its production expenses within six months. "In 24 Weeks 'The Ramblers' Managed to Get 'Off the Nut,'" n.s., n.d., n.p., *The Ramblers* clippings file, BRTC.

94. Dunley, "The Ramblers."

95. The notion of "affordances" is derived from Ian Hutchby's approach to medium-centered criticism. According to Hutchby, "Affordances are functional and relational aspects [of media] which frame, while not determining, the possibilities for agentic action in relation to an object." As such, affordances open a "third way between the [constructivist] emphasis on the shaping power of human agency and the [realist] emphasis on the shaping power of technical capacities." Ian Hutchby, "Technologies, Texts, and Affordances," *Sociology* 35, no. 2 (2001): 444.

96. On this aspect of Burns and Allen's sound shorts, see Charles Wolfe, "'Cross-Talk': Language, Space, and the Burns and Allen Comedy Film Short," *Film History* 23, no. 3 (2011): 300–312.

97. Release dates for Clark and McCullough's Fox shorts are difficult to pinpoint. The trade press offers exact dates for only two, *The Bath Between* and *The Diplomats*, which are listed in *Motion Picture News*'s regular "Complete Release Schedule" column as having both been released on February 17, 1929. *Film Daily* lists seven—*Belle of Samoa, Beneath the Law, In Holland, Knights Out, The Medicine Men, The Music Fiends,* and *Waltzing Around*—as "Releases for First Three Months of '29" ("What the Field Has to Offer in Shorts," March 31, 1929, 20), but does not give further specifics. It is possible that these seven were made simultaneously available to Fox exhibitors.

98. Edwin M. Bradley, *The First Hollywood Sound Shorts, 1926–1931* (Jefferson, NC: McFarland, 2005), 47.

99. "Profiles, III," 40.

100. "Bath Between," August 30, 1928, 20th Century–Fox Script Collection, USC. Clark and McCullough's *Music Box* routine involved two hotel rooms, a connecting bathroom, and the risqué encounters between the duo, who occupy one room, and a beautiful wife and her violent husband, in the other. The skit soon became a classic in the burlesque playbook, performed by subsequent burlesque double acts like Smith and Dale. On the hotel scenes of burlesque, see Davis, *Baggy Pants Comedy*, 215–20.

101. On con games and courtroom sketches in burlesque, see Davis, *Baggy Pants Comedy*, ch. 10 and 228–33.

102. "Profiles, III," 42. Extant script materials at the New York State Archives show that Clark here misremembered the film's ending: the film in fact closes on the spectacle of the police trying to break into the judge's chamber while, within, we hear the judge and Clark and McCullough express enjoyment of their private dance ("Oh, boy, ain't we got fun!" etc.). Dialogue script, *Beneath the Law* ("dialogue as taken from screen"), undated, New York State Archives.

103. Gilles Deleuze, *The Logic of Sense,* trans. Mark Lester (New York: Columbia University Press, 1990 [1969]), 59. Something of this approach is forecast in one of Clark and McCullough's earlier Fox shorts, *The Music Fiends* (ca. February 1929), whose plot has the boys mistakenly hired as musicians for a society lady's swanky party. In a lengthy routine as they prepare their recital (four script pages out of a total of twenty-one), the duo persuades their host to move the piano to different spots around the room in an effort to find the "perfect" lighting (because too bright light "hurts [McCullough's] ears"), causing all manner of furniture destruction and physical inconvenience. (By the time the recital starts, the piano is jutting out of a broken window with the host propping

up the instrument's leg base.) As would become more prominent at RKO, the routine is structured as a game whose conditions of play are governed only by the whim of the players: "Now where do you want it?" Clark repeatedly asks, to which McCullough responds in turn, "Turn it around and I'll show you," indicating a new location each time. Dialogue script, *The Music Fiends* ("dialogue as taken from the screen"), undated, New York State Archives.

104. Ben Holmes, "Clark and McCullough #3," final script, June 19, 1933, 4, RKO.

105. Ibid., 5.

106. Ibid., 6.

107. Gilles Deleuze, "Bartleby; or, The Formula," in *Essays Critical and Clinical*, trans. Daniel W. Smith and Michael A. Greco (London: Verso, 1999), 82.

108. Ben Holmes, "Clark and McCullough #4," story outline, June 29, 1933, 1–2, RKO.

109. Ibid., 6.

110. Ben Holmes, "Clark and McCullough #4," final script, July 8, 1933, 47–48, RKO.

111. Jenkins, *What Made Pistachio Nuts?*, 130–31.

112. See Jacques Rancière's discussion of wit in *Mute Speech,* trans. James Swenson (New York: Columbia University Press, 2011), 153–54.

113. "Profiles, III," 40.

114. "Profiles—Comedian," 36.

115. David N. Bruskin, *The White Brothers: Jack, Jules, & Sam* White (Metuchen, NJ: Scarecrow Press, 1990), 300.

116. Some qualification is called for here since even their earlier shorts had occasionally fallen back on some of film slapstick's most hackneyed plots—as in, for instance, the Fox two-reeler *Detectives Wanted* (July 1929), which follows the "haunted house" template popularized by Harold Lloyd's *Haunted Spooks* (March 1920). The distinction I am drawing lies rather at the level of physical performance and comic "business," which begin to take on a fairly standard knockabout sheen in Clark and McCullough's later work.

117. Steve Massa, "Alibi Bye Bye," in *Slapsticon '08* program notes (2008), 50.

118. Nic Sammond, *Birth of an Industry: Blackface Minstrelsy and the Rise of American Animation* (Durham, NC: Duke University Press, 2015), 121.

119. See Wolfe, "'Cross-Talk.'"

120. "The Social Life of the Newt" is from Benchley's *Of All Things* (New York: Henry Holt, 1921); "Do Insects Think?" and "Polyp with a Past" from his *Love Conquers All.* Benchley's shorts are discussed at greater length in the next chapter.

121. Arthur Pollock, "The Theater," *Brooklyn Daily Eagle*, April 9, 1929, n.p., *A Night in Venice* clippings file, BRTC.

122. It was common practice in the Healy/Stooge MGM shorts to make use of musical numbers scrapped from feature films. The "The Turn of the Fan" sequence from *Nertsery Rhymes* (July 1933), for example, was lifted from the uncompleted 1930 musical *The March of Time;* a choral dance from *The Big Idea* was abandoned from the Joan Crawford musical *The Dancing Lady* (1933); and the "I'm Sailing on a Sunbeam" song from *Hello Pop!* (September 1933) recycled from *It's a Great Life* (1929).

123. Bourdieu, *The Rules of Art,* 206.

124. See Kathryn Fuller-Seeley, "Becoming Benny: Evolution of Jack Benny's Character Comedy from Vaudeville to Radio," *Studies in American Humor,* series 4, vol. 1, no. 2 (2015): 163–91.

125. "Robert Benchley," *Spot,* April 1941, 11.

CHAPTER 2. "THE STIGMA OF SLAPSTICK"

1. On vaudeville performers' direct interaction with live audiences, see Henry Jenkins, *What Made Pistachio Nuts? Early Sound Comedy and the Vaudeville Aesthetic* (New York: Columbia University Press, 1992), 73–77.

2. Charles Wolfe, "Vitaphone Shorts and *The Jazz Singer,*" *Wide Angle* 12, no. 3 (July 1990): 64.

3. Michael Warner, *Publics and Counterpublics* (New York: Zone Books, 2005), 67, 74 (emphasis in original).

4. As numerous social and cultural historians have argued, blackness provided a symbol of difference that allowed turn-of-the-century publics to subsume previous ethnic divisions beneath a kind of ersatz mass-cultural whiteness. See, for example, Noel Ignatiev, *How the Irish Became White* (New York: Routledge, 1996); David Nasaw, *Going Out: The Rise and Fall of Public Amusements* (New York: Basic Books, 1993); David Roediger, *The Wages of Whiteness: Race and the Making of the American Working Class* (London: Verso, 1991); and Michael Rogin, *Blackface/White Noise: Jewish Immigrants in the Hollywood Melting Pot* (Berkeley: University of California Press, 1996). African Americans thus constituted what Anna Everett calls a "phantom audience for Hollywood's czars," or, as Catherine Jurca adds, "a market to be exploited rather than a public to be solicited." Anna Everett, *Returning the Gaze: A Genealogy of Black Film Criticism, 1909–1949* (Durham, NC: Duke University Press, 2001), 193; Catherine Jurca, *Hollywood 1938: Motion Pictures' Greatest Year* (Berkeley: University of California Press, 2012), 70.

5. As Margaret Thorp summarized these themes: "The movies are furnishing the nation with a common body of knowledge. What the classics once were in that respect, what the Bible once was, the cinema has become for the average man. Here are stories, names, phrases, points of view which are common national property. The man in Cedar Creek, Maine, and the man in Cedar Creek, Oregon, see the same movie in the same week.... The movies span geographic frontiers; they give the old something to talk about with the young; they crumble the barriers between people of different educations and different economic backgrounds." Thorp, *America at the Movies* (New Haven, CT: Yale University Press, 1939), 271–72.

6. Catherine Jurca, "Motion Pictures' Greatest Year (1938): Public Relations and the American Film Industry," *Film History* 20, no. 3 (2008): 344–56; and Jurca, *Hollywood 1938*.

7. Peter Vischer, "Broadway," *EHW,* March 15, 1930, 18.

8. Pierre Bourdieu, *The Rules of Art: Genesis and Structure of the Literary Field,* trans. Susan Emanuel (Stanford, CA: Stanford University Press, 1995), 214.

9. Hays's remarks are reproduced in Will H. Hays, *See and Hear: A Brief History of Motion Pictures and the Development of Sound* (New York: MPPDA, 1929), 48–49.

10. Susan Douglas details how radio made the enjoyment of classical music the province of listeners who otherwise had limited access to live recitals and performances. Susan J. Douglas, *Listening In: Radio and the American Imagination* (New York: Random House, 1999), 150. On radio and musical uplift, see also Michele Hilmes, *Radio Voices: American Broadcasting, 1922–1952* (Minneapolis: University of Minnesota Press, 1997), ch. 1; and Steve J. Wurtzler, *Electric Sounds: Technological Change and the Rise of Corporate Mass Media* (New York: Columbia University Press, 2007), ch. 4.

11. Vitaphone promotional booklet, undated, ca. 1926, WB.

12. Vitaphone Corporation to John T. Adams, September 14, 1926, WB.

13. Wolfe, "Vitaphone Shorts and *The Jazz Singer*," 62. On the shift away from opera, see also Jennifer Fleeger, *Sounding American: Hollywood, Opera, and Jazz* (New York: Oxford University Press, 2014), 39–43.

14. See in particular Lewis Erenberg, *Steppin' Out: New York Nightlife and the Transformation of American Culture, 1890–1930* (Westport, CT: Greenwood Press, 1981).

15. Advertisement for Vitaphone, *Motion Picture Classic,* December 1929, 10, quoted in Jenkins, *What Made Pistachio Nuts?,* 160.

16. Fitzgerald, "Echoes of the Jazz Age" (1931), in *The Crack-Up* (New York: New Directions, 1945), 14. On New York's emerging cultural leadership during this period, see ch. 1.

17. On earlier efforts to harness cinema to the project of cultural dissemination, see William Uricchio and Roberta E. Pearson, *Reframing Culture: The Case of the Vitagraph Quality Films* (Princeton, NJ: Princeton University Press, 1993), esp. chs. 1 and 2; and my essay, "'Made for the Masses with an Appeal to the Classes': The Triangle Film Corporation and the Failure of Highbrow Film Culture," *Cinema Journal* 44, no. 2 (Winter 2005): 3–33.

18. Quotes taken from Joseph Medill Patterson, "The Nickelodeons: The Poor Man's Elementary Course in the Drama," *Saturday Evening Post,* November 23, 1907, 38; advertisement for Vitaphone, *Motion Picture Classic.*

19. Advertisement for Vitaphone, *Photoplay,* October 1929, 153.

20. Paul Seale, "'A Host of Others': Towards a Nonlinear History of Poverty Row and the Coming of Sound," *Wide Angle* 13, no. 1 (January 1991): 72–103.

21. Fitzhugh Green, *The Film Finds Its Tongue* (New York: G. B. Putnam's Sons, 1929), 87. According to Green, the appeal of operatic and classical shorts was further hampered by regional taste cultures, a problem that became especially marked after 1928, when Warners began to allow exhibitors to choose freely from the catalog of Vitaphone reels. "Talkie shows [now] had to be booked more like vaudeville than like pictures . . . Some cities liked the opera numbers; others wouldn't stand for them. Western and southern states in particular objected to highbrow stuff. Those first numbers that had been necessary to launch the thing in New York did not go over so well in the back country" (83).

22. "Schenck Outlines Complete MGM Plan," *EHMPW,* September 15, 1928, 28; "Julius Singer Names Sales Head for Universal Shorts in Sound," *EHMPW,* September 15, 1928, 35; Paramount advertisement, *EHMPW,* December 15, 1928, insert.

23. Wolfe, "Vitaphone Shorts and *The Jazz Singer*," 66.

24. "Short Talking Comedies Are Rapidly Becoming Favorites," press sheet for *The Right Bed* (April 1929), 2, EPS. The Coronet shorts are discussed at greater length in the following chapter.

25. "32 Talking Films of Christie Firm to Be Prereleased," *EHMPW,* December 15, 1928, 36; "January Brings Christie Talking Plays and Cohen Negro Stories," *EHW,* January 5, 1929, 44.

26. "Sound Places Shorts on Feature Plane, Says Al Christie," *EHW,* June 15, 1929, 128.

27. On *Yamekraw,* see Fleeger, *Sounding American,* 43–52. The Darktown shorts were not the only black-cast short-subject series from the conversion period: in the summer of 1929, Pathé produced six two-reel comedies starring the vaudeville team of Buck and Bubbles (Ford Lee Washington and John William Sublett), released in the 1929–1930 season.

28. "January Brings Christie Talking Plays," 44; George Kent Shuler, "Pictures and Personalities," *Motion Picture Classic*, April 1929, 17.

29. Richard Leacock, *Humor: Its Theory and Technique* (Toronto: Dodd, Mead, 1935). 85.

30. "Opinions on Pictures," *MPN*, May 11, 1929, 1646.

31. Bourdieu, *The Rules of Art*, 254.

32. "Speaking Briefly of Comedy," *EHW*, February 9, 1929, 40.

33. Green, *The Film Finds Its Tongue*, 282.

34. On the "low other," see Peter Stallybrass and Allon White, *The Poetics and Politics of Transgression* (Ithaca, NY: Cornell University Press, 1986), esp. introduction.

35. Warners' own stockholders' reports give an interesting perspective on the shift. Up until 1930, those reports regularly included a list of the "outstanding artists of the screen and of the operatic, legitimate and vaudeville stage appearing in 'Vitaphone Varieties' of short subjects"; but in the 1931 report, any reference to "outstanding artists" was replaced by a straightforward listing of series. See the annual reports dated August 30, 1930, and August 29, 1931, WB.

36. The Roscoe "Fatty" Arbuckle scandal, in which the comedian was tried for manslaughter following the death of Virginia Rappe at a wild party, has been given book-length treatments by David Yallop, *The Day the Laughter Stopped* (New York: St. Martin's Press, 1976); and Andy Edmonds, *Frame Up! The Untold Story of Roscoe "Fatty" Arbuckle* (New York: William Morrow, 1991). Few discussions, however, are as well grounded in cultural history as Hilary A. Hallett's, in *Go West, Young Women! The Rise of Early Hollywood* (Berkeley: University of California Press, 2013), ch. 5; and Gary Alan Fine's, in *Difficult Reputations: Collective Memories of the Evil, Inept, and Controversial* (Chicago: University of Chicago Press, 2001), ch. 4.

37. "'Times Have Changed but Not Comedy,' Says Arbuckle," press sheet for *Hey Pop* (August 1932), 2, BRTC. Worth noting, too, is Arbuckle's recycling of plot situations from earlier comedies, both his own and other comedians'—the way, for instance, that the grocery store scenes of *How've You Bean* (June 1933) replay the opening of Arbuckle's debut Comique short, *The Butcher Boy* (April 1917), or how the bomb-in-a-cake narrative of *In the Dough* (November 1933) revisits the plot of Charlie Chaplin's *Dough and Dynamite* (October 1914).

38. Lizabeth Cohen, *A Consumers' Republic: The Politics of Mass Consumption in Postwar America* (New York: Vintage Books, 2003), 18.

39. "Broadway and/or United States," *MPH*, June 3, 1933, 19.

40. "Thus, although the culture industry undeniably speculates on the conscious and unconscious state of the millions towards which it is directed, the masses are not primary, but secondary, they are an object of calculation; an appendage of the machinery. The customer is not king, as the culture industry would have us believe, not its subject but its object." Theodor Adorno, "Culture Industry Reconsidered," in *The Culture Industry*, ed. J. M. Bernstein (New York: Routledge, 2002), 99.

41. Richard Koszarski, *An Evening's Entertainment: The Age of the Silent Feature Picture, 1915–1928* (Berkeley: University of California Press, 1990), 9.

42. In his study of picture palace impresario Samuel "Roxy" Rothafel, Ross Melnick proposes the concept of the "unitary text" to define the silent-era moviegoing experience as a "collective textual event" that was "authored" by the theater manager. See Melnick, *American Showman: Samuel "Roxy" Rothafel and the Birth of the Entertainment Industry* (New York: Columbia University Press, 2012), 8–23.

43. "Audien Shorts Will Kill Presentations, Declares Jack White," *EHW*, April 27, 1929, 36.

44. Martin J. Quigley, editorial, *EHW*, May 25, 1929, 18.

45. Siegfried Kracauer, "The Cult of Distraction," in *The Mass Ornament*, ed. and trans. Thomas Levin (Cambridge, MA: Harvard University Press, 1995), 232–328.

46. On these points, see Nic Sammond, "A Space Apart: Animation and the Spatial Politics of Conversion," *Film History* 23, no. 3 (2011): 269–72.

47. For example, the chain prologue company Fanchon and Marco continued to supply movie theaters with prepackaged stage shows until shutting down in 1937. See Phil Wagner, "'An America Not Quite Mechanized': Fanchon and Marco, Inc. Perform Modernity," *Film History* 23, no. 3 (2011): 251–67. On live acts in rural theaters, see Gregory A. Waller, "Hillbilly Music and Will Rogers: Small Town Picture Shows in the 1930s," in Melvyn Stokes and Richard Maltby, eds., *American Movie Audiences: From the Turn of the Century to the Early Sound Era* (London: BFI, 1999), 164–79.

48. "Roxy Turns Universal Sales Convention into Open Forum," *Universal Weekly*, June 7, 1930, 19, quoted in Melnick, *American Showman*, 348.

49. Lewis, "Program Fillers or Program Builders?," *MPH*, March 14, 1931, 75, quoted in Thomas Doherty, "This Is Where We Came In: The Audible Screen and the Voluble Audience of Early Sound Cinema," in Stokes and Maltby, eds., *American Movie Audiences*, 149.

50. E. A. Schiller, "Shorts Solve Showman's Problem of 1931," *MPH*, March 14, 1931, 60.

51. Terry Ramsaye, "The Short Picture—Cocktail of the Program," *MPH*, March 14, 1931, 59.

52. Advertisement for MGM shorts, *EHW*, June 29, 1929, insert.

53. On the Bobby Jones golf shorts, see Harper Cossar, "Bobby Jones, Warner Bros., and the Short Instructional Film," in Janet Staiger and Sabine Hake, eds., *Convergence, Media, History* (New York: Routledge, 2009), 102–11.

54. A noted authority on bridge, Milton Work entered into a contract with Vitaphone in the fall of 1929 to appear in instructional bridge shorts. The idea to use domestic comedy as a framework seems to have been his own. See Milton C. Work to Nathan Vidaver, August 29, 1929, WB.

55. Descriptions quoted from advertisement for MGM shorts, *MPH*, May 16, 1931, 20–21; and "Bruce with Educational," *MPH*, May 27, 1933, 50.

56. In 1934, a *Film Daily* survey ranked animated cartoons first among audience preferences, and live-action comedy shorts third. "Critics Forum," *Film Daily*, May 7, 1934, 1. Five years later, one of the earliest film surveys conducted by George Gallup's Audience Research Institute revealed that "cartoon comedies" remained the best-liked short subjects, with live-action comedy ranking dead last. Audience Research Institute, "The Verdict of the Box-Office," October 1941, in the microfilm *Gallup Looks at the Movies: Audience Research Reports 1940–1950* (Wilmington, DE: Scholarly Resources, 1979).

57. See Paige Reynolds, "'Something for Nothing': Bank Night and the Refashioning of the American Dream," in Kathryn H. Fuller-Seeley, *Hollywood in the Neighborhood: Historical Case Studies of Local Moviegoing* (Berkeley: University of California Press, 2008), 208–30.

58. Gary D. Rhodes, "'The Double Feature Evil': Efforts to Eliminate the American Dual Bill," *Film History* 23, no. 1 (2011): 59.

59. "The Hollywood Scene," *MPH*, October 5, 1935, 57.

60. "Most Trade Leaders Denounce Double Featuring as a Menace," *MPH*, November 21, 1931, 32.

61. RKO budget data has been averaged out from the many short-subject production files held at the RKO Studio Collection, Performing Arts Special Collections, University of California, Los Angeles; Columbia budget statistics from "Lists—Costs," JWC.

62. "The Hollywood Scene," 57.

63. "76 Per Cent of Patrons Ask for Single Bills and Shorts," *MPH*, April 2, 1932, 25.

64. Ibid.

65. "Public Protests Double Features; Suggests Shorts as a Stimulant," *MPH*, March 25, 1933, 11. See also the correspondence between Lew Maren and Thomas Gerety, dated March 2 and 18, 1933, Files—1930s, HRC.

66. "How Public Leaders View Double Feature Showings," *MPH*, May 13, 1933, 50.

67. "Public Protests Double Features," 11.

68. "How Public Leaders View Double Feature Showing," 50; "Double Bills Lose by 5 to 1 in Survey," *MPH*, June 9, 1934, 18.

69. "Duals Ruling Called Aid to Little Fellow," *Motion Picture Daily*, August 17, 1934, 1, 6, quoted in Rhodes, "'The Double Feature Evil,'" 61.

70. For the relevant portion of the Motion Picture Code, see "Complete Text of the Code," *MPH*, December 2, 1933, 32.

71. "Philly Dual Bill Case Continued to May 17," *Film Daily*, May 9, 1934, 2.

72. Frank Ricketson, Jr., *The Management of Motion Picture Theatres* (New York: McGraw-Hill, 1938), 79, quoted in Rhodes, "The Double Feature Evil," 66.

73. The "moral collapse" quote is from Bosley Crowther, "Two-Reeler's Comeback," *NYT*, October 26, 1941, SM19.

74. Douglas W. Churchill, "Hollywood Rediscovers the 'Short,'" *NYT*, July 18, 1937, X3.

75. Quote taken from Herman Boxer, Temporary Complete Dialogue Continuity, "Return to Life," August 5, 1938, 19 pp., MGM Shorts Collection, AMPAS. The *Soldiers of Peace* and *What Do You Think?* films were released through the "MGM Miniatures" line.

76. "A New Deal for Shorts," *MGM Shortstory*, May–June 1939, 10.

77. Allen Saunders, "A Few Short Words," *MGM Shortstory*, August–September 1938, 5.

78. The concept of useful film, Wasson and Acland explain, "overlaps with, but is not equivalent to, similar terms such as 'functional film,' 'educational film,' 'non-fictional film,' and 'non-theatrical film.' We define useful cinema to include experimental films and a variety of didactic films that are fictional as well as non-fictional, narrative as well as non-narrative. The concept of useful cinema does not so much name a mode of production, a genre, or an exhibition venue as it identifies a disposition, an outlook, and an approach toward a medium on the part of institutions and institutional agents." It is thus not a historically specific designation but a category that encompasses a range of functional conceptions of cinema as a pedagogical medium. Wasson and Acland, "Introduction: Utility and Cinema," in Wasson and Acland, eds., *Useful Cinema* (Durham, NC: Duke University Press, 2011), 4.

79. Robert W. Desmond, "Something to Think About," *Christian Science Monitor*, August 17, 1938, 4, 14.

80. Halsey Raines, "Shorts as a School Subject," *MGM Shortstory*, January–February 1938, 9.

81. See, for example, Eric Smoodin, "'What a Power for Education!' The Cinema and Sites of Learning in the 1930s," and Charles R. Acland, "Hollywood's Educators: Mark

May and Teaching Film Custodians," in Wasson and Acland, eds., *Useful Cinema*, 18–33 and 59–80; Craig Kridel, "Educational Film Projects of the 1930s: Secrets of Success and the Human Relations Film Series," in Devin Orgeron, Marsha Orgeron, and Dan Streible, eds., *Learning with the Lights Off: Educational Film in the United States* (New York: Oxford University Press, 2012), 215–29.

82. One very proximate impetus here may have been a desire to counteract the bad press brought by the publication in 1933 of Henry James Forman's *Our Movie Made Children*, an alarmist condensation of a series of Payne Fund Studies on motion pictures' influence on America's youth. On the Payne Fund Studies and their ensuing controversy, see Garth S. Jowett, Ian C. Jarvie, and Kathryn H. Fuller, *Children and the Movies: Media Influence and the Payne Fund Controversy* (Cambridge: Cambridge University Press, 1996).

83. Secrets of Success Manual, quoted in Kridel, "Educational Film Projects of the 1930s," 216. The *Secrets of Success* films were composed of excerpted scenes from noncurrent commercial features.

84. Alice Keliher, "The Short Way to Education," *MGM Shortstory*, June 1938, 16.

85. Ibid.; "Short Waves of Response," *MGM Shortstory*, November–December 1939, 4.

86. Keliher, "The Short Way," 16.

87. Marguerite Tazelaar, "Evolution of the Cinema Shorts," *NYHT*, January 30, 1938, VI.6.

88. See Keliher, "The Short Way"; Marilla Waite Freeman, "Short Way to the Library," *MGM Shortstory*, April 1938, 6–7; "Short Waves of Response," *MGM Shortstory*, July–August, 1939, 4; Edgar J. Hoover [sic], "Combating Crime through Movies!" *MGM Shortstory*, July–August 1939, 6–7.

89. "1939—a Short Year," *MGM Shortstory*, August–September 1938, 7.

90. "Farewell to the Custard Pie," *MGM Shortstory*, November 1937, 7.

91. Acland, "Hollywood's Educators," 70.

92. Advertisement, Independent Theatre Owners' Association, *Hollywood Reporter*, May 3, 1938, 5; John C. Flinn, "Pix Slipping in Stix," *Variety*, July 20, 1938, 1; Oscar A. Doob, "It's a State of Mind Says Loew's Advertising Head," *MPH*, July 16, 1938, 87, all quoted in Jurca, "Motion Pictures' Greatest Year (1938)," 344–45.

93. Jurca, "Motion Pictures' Greatest Year (1938)," 350–52. See also Jurca, *Hollywood 1938*, ch. 3, for a more detailed examination of the "Motion Pictures' Greatest Year" campaign advertising.

94. Jurca, *Hollywood 1938*, 101–12. The phrase "idols of consumption" is from Leo Lowenthal, *Literature and the Image of Man* (Boston: Beacon Press, 1957), quoted in Richard Dyer, *Stars*, 2nd ed. (London: BFI, 1998), 39. Jurca's key example of this recalibration of stardom is Myrna Loy, who, despite starting her career playing "exotics and sirens" in silent films, was, by the end of the 1930s, promoted as a "natural girl" who represented the "sweet virtues of normalcy." Jurca, *Hollywood 1938*, 109.

95. Phil Wagner, "'A Particularly Effective Argument': *Land of Liberty* (1939) and the Hollywood Image (Crisis)," *Film & History* 41, no. 1 (Spring 2011): 9.

96. The terms "logic of difference" and "logic of equivalence" are derived from Ernesto Laclau, for whom they describe opposed poles of social and political structuration. See his essay, "Articulation and the Limits of Metaphor," in Laclau, *The Rhetorical Foundations of Society* (London: Verso, 2014), 53–78.

97. Crowther, "Two-Reeler's Comeback," SM8, 19.

98. Advertisement for Universal, *MPH*, July 11, 1936, 97–100.

99. Crowther, "Two-Reeler's Comeback," SM19.

100. The MGM series was interrupted when Benchley briefly switched his allegiances to Paramount, where he appeared in nine short subjects between 1940 and 1942.

101. *Radio Daily* poll cited in Wes D. Gehring, *"Mr. B" or Comforting Thoughts about the Bison: A Critical Biography of Robert Benchley* (Westport, CT: Greenwood Press, 1992), 125.

102. Tazelaar, "Evolution of the Cinema Shorts," 6.

103. "A Humorist Comes Down to Earth," *New York Sun,* February 4, 1939, n.p., RBS.

104. Warren Susman, "The Culture of the Thirties," in *Culture as History: The Transformation of American Society in the Twentieth Century* (New York: Pantheon Books, 1984), 156.

105. On "averageness" and Depression-era social sciences, see Sarah E. Igo, *Averaged Americans: Surveys, Citizens, and the Making of a Mass Public* (Cambridge, MA: Harvard University Press, 2007), chs. 1–3.

106. Benchley quoted in "The Films of Robert Benchley," *Film Fan Monthly* 59 (May 1966): 3.

107. "Mr. Benchley Laughs Out Loud," *Jacksonville Times-Union,* December 2, 1928, n.p., RBS.

108. *Screenland,* November 1928, n.p., RBS.

109. *Variety,* June 27, 1928, quoted in Edwin M. Bradley, *The First Hollywood Sound Shorts, 1926–1931* (Jefferson, NC: McFarland, 2005), 18.

110. Robert Benchley, "A Possible Revolution in Hollywood," *Yale Review* (Autumn 1931): 101, 103. Benchley's celebrity as the leading figure of urbane East Coast humor in fact bespoke a significant contradiction in Hollywood's own cultural position taking during the early sound period, insofar as the film industry was seeking to appropriate the cachet of a sensibility that itself held the industry in considerable disdain. The best-known instance of that contradiction is unquestionably provided in the career of Ben Hecht, who once described Hollywood in the pages of the *New Yorker* as "the Waterloo of America's mental progress." Hecht was persuaded to work for the industry, however, when fellow *New Yorker* contributor Herman Mankiewicz famously wrote him to explain that "millions are to be grabbed out here and your only competition is idiots." Ben Hecht, "America's Waterloo," *New Yorker,* July 18, 1925, 6.

111. Vivian Shaw, "The Cuckoo School of Humour in America," *Vanity Fair,* May 1924, 46.

112. Robert E. Rogers, Lecture Outline VIII, "Literature for Recreation, Part II: The Literature of Wit and Humor," Division of University Extension, Massachusetts Department of Education (1932–1933), RBS. On the absurdist quality of Benchley's writings, see ch. 1.

113. Benchley's later short, *How to Vote* (September 1936), reuses material from this speech.

114. Norris W. Yates, *Robert Benchley* (New York: Twayne, 1968), 95–96. One could also add other notable postbellum humorists who worked in this vein, such as George Horatio Derby (in his persona as John Phoenix), Charles Farrar Browne (as Artemus Ward), and even Samuel Clemens (as Mark Twain).

115. Michel Foucault, *The Order of Things: An Archaeology of the Human Sciences* (New York: Vintage Books, 1994 [1966]), xv.

116. Ibid., xvi–xvii.

117. The dichotomy of "good taste" versus "mass taste" I derive from Jennifer Lynn Peterson, *Education in the School of Dreams: Travelogues and Early Nonfiction Film* (Durham, NC: Duke University Press, 2013), 112–16.

118. "A Fifteen-Year Debut," *NYHT,* October 27, 1940, 3.

119. "1939, A Short Year," *MGM Shortstory,* August–September 1938, 7.

120. "Robert Benchley," *Spot,* April 1941, 11.

121. "A Humorist Comes Down to Earth," n.p. (emphasis added).

122. Sianne Ngai, *Our Aesthetic Categories: Zany, Cute, Interesting* (Boston, MA: Harvard University Press, 2012), ch. 3.

123. "All Aboard for Dementia Praecox," *New York American,* June 18, 1934.

124. Ngai, *Our Aesthetic Categories,* 188–89, 203–8. The concept of a "becoming-woman" of labor is derived from Antonella Corsani, "Beyond the Myth of Woman: The Becoming-Transfeminist of (Post-)Marxism," *SubStance* 112 (2007): 107–38.

125. Ngai, *Our Aesthetic Categories,* 188.

126. Parsons cited in Bill Cassara, *Edgar Kennedy: Master of the Slow Burn* (Boalsburg, PA: Bear Manor Media, 2005), 124, 128.

127. Umberto Eco, "The Frames of Comic Freedom," in Thomas Sebeok, ed., *Carnival!* (Berlin: Mouton, 1985), 1–9.

128. Tazelaar, "Evolution of the Cinema Shorts," 6.

129. "Meet Mr. Benchley," *Listeners Digest,* April 1939, 67, Robert Benchley clippings file, AMPAS.

130. "A Humorist Comes Down to Earth," n.p.

131. Raymond Williams, *Culture and Society: 1780–1950* (New York: Columbia University Press, 1958), 300.

CHAPTER 3. "THE SPICE OF THE PROGRAM"

1. Jack White quoted in David Bruskin, *The White Brothers: Jack, Jules, & Sam White* (Metuchen, NJ: Scarecrow Press, 1990), 116.

2. Earle W. Hammons in Joseph P. Kennedy, ed., *The Story of the Films* (Chicago: A. W. Shaw, 1927), 151.

3. "Educational Films Branching Out," *Moving Picture World,* August 31, 1918, 1244.

4. The name "Mermaid Comedies" seems to have originated in reference to the bathing beauties in one of White's early comedies for Educational. "I went to Balboa and made a picture about bathing beauties," White explained. "New York chose to call the entire series Mermaids. They didn't ask me; they just went ahead and made the main title 'Mermaid Comedy.'" White in Bruskin, *The White Brothers,* 73.

5. Richard M. Roberts, "'Mixed Nuts' and Educational Pictures," *Classic Images,* no. 211 (January 1992), 26. Budget figures taken from Bruskin, *The White Brothers,* 90, 104.

6. Theater statistics taken from Bruskin, *The White Brothers,* 122; Hammons quoted in Kennedy, ed., *The Story of the Films,* 151. Sennett was let go from Pathé during Joseph P. Kennedy's cost-cutting reorganization of the company; see Cari Beauchamp, *Joseph P. Kennedy Presents: His Hollywood Years* (New York: Knopf, 2009), ch. 13.

7. "What the Picture Did for Me," *MPH,* December 15, 1934, 63; August 18, 1934, 58; and January 19, 1935, 87.

8. Leonard Maltin, *The Great Movie Shorts* (New York: Bonanza Books, 1972), 12.

9. James Agee, "Comedy's Greatest Era" (1949), reprinted in *Agee on Film,* vol. 1 (New York: Perigee Books, 1958), 4.

10. See my introduction.

11. Gladys Hall, "Charlie Chaplin Attacks the Talkies," *Motion Picture Magazine,* May 1929, 29, quoted in Kenneth S. Lynn, *Charlie Chaplin and His Times* (London: Aurum Press, 1997), 321.

12. Walter Kerr, *The Silent Clowns* (New York: DaCapo Press, 1975), 26.

13. Quoted in Bruskin, *The White Brothers,* 129.

14. "Vocafilm Premiere Fails," *EHMPW,* August 5, 1927, 388; "Educational Acquires Vocafilm," *EHMPW,* January 14, 1928, 121. For more on Vocafilm, see Douglas Gomery, *The Coming of Sound* (New York: Routledge, 2005), 82–85.

15. Quoted in Bruskin, *The White Brothers,* 130.

16. "Subtle Humor in New Series of Talking Films," press sheet for *Prince Gabby* (1929), 3, EPS.

17. "Edward Everett Horton Has Star Role in Talking Film" and "Talking Pictures Make Horton Big Comedy Favorite," press sheet for *Ask Dad* (1929), 2, EPS. Although Horton is today best remembered for the "sissy" overtones of his work in Depression-era features, his Coronet shorts instead generally cast him as something of a ladies' man who, variously, steals the affections of the secretary his son intends to marry in *Ask Dad,* commences a dalliance with a young woman before his divorce has been finalized in *The Right Bed* (April 1929), and writes love sonnets to other men's spouses in *Trusting Wives* (June 1929). The apparent exception here, interestingly, is the very first in the series, *The Eligible Mr. Bangs* (January 1929), which hints at the queer dimension to his later persona by casting Horton as a "girl-hater" whose "indifference to unmarried girls . . . [is] caused by his fear of being trapped into marriage." "The Story," press sheet for *The Eligible Mr. Bangs* (1929), 1, EPS. In general, queerness never achieved the (admittedly coded) visibility in shorts that it did in feature films in the early sound period, even though many of the character actors most associated with sissy or pansy roles in features had first passed through sound shorts (e.g., Horton and Franklin Pangborn at Educational; Grady Sutton at Roach; Eric Blore in presentation acts at Vitaphone and MGM). On queer representation in Depression-era Hollywood, see David M. Lugowski, "Queering the (New) Deal: Lesbian and Gay Representation and the Depression-Era Cultural Politics of Hollywood's Production Code," *Cinema Journal* 38, no. 2 (Winter 1999): 3–35.

18. Statistics on average shot lengths taken from Charles O'Brien, "Multiple Language Versions and National Films, 1930–1933: Statistical Analysis, Part I," *Cinéma et Cie* 6 (Winter 2005): 45–52. See also David Bordwell, "The Introduction of Sound," in Bordwell, Janet Staiger, and Kristin Thompson, *The Classical Hollywood Cinema: Film Style and Mode of Production to 1960* (New York: Columbia University Press, 1985), 304.

19. "Voice of the Industry," *EHW,* December 27, 1930, 47; "A Showman Discusses the Short-Comings of the Short Feature!" *MPH,* April 23, 1932, 49.

20. Lea Jacobs, *Film Rhythm after Sound: Technology, Music, and Performance* (Berkeley: University of California Press, 2015), 3.

21. Michel Chion, *Audio-Vision: Sound on Screen* (New York: Columbia University Press, 1994), 16–17. For a full discussion of this issue, see Kevin Brownlow, "Silent Films: What Was the Right Speed?" *Sight & Sound* (Summer 1980): 164–67; also Kerr, *The Silent Clowns,* 32–38.

22. "Jack White's Rare Comedy Genius Is in Evidence in Latest Talking Picture," press sheet for *Cold Shivers* (1929), 3, EPS.

23. This approach is perhaps most famously associated with René Clair's distinction between the film *sonore* and the film *parlant.* René Clair, "Talkie versus Talkie" (1929), in Richard Abel, ed., *French Film Theory and Criticism: A History/Anthology, 1907–1939* (Princeton, NJ: Princeton University Press, 1993), 39–40.

24. "Jack White Uncovers New Entertainment in Talkies," press sheet for *Hunting the Hunter* (1929), 2, EPS.

25. Jack White quoted in Bruskin, *The White Brothers*, 132.

26. Sennett had contributed uncredited directorial chores on a number of his productions in the 1920s and had helmed in its entirety the eight-reel feature, *The Good-Bye Kiss*, in late 1927 (distributed through First National the following year). He had not, however, performed regular directing duties since 1914.

27. "Mack Sennett Sees Sounds as Big Help to Short Comedies," press sheet for *The Bees' Buzz* (1929), 2, EPS.

28. "Sound Enhances Effectiveness of Fun in Mack Sennett Film," press sheet for *The Bees' Buzz*, 3.

29. "Peace and Quiet," 1, *The Lion's Roar*, Production Files, MSC.

30. The notion of "functional equivalence" derives from David Bordwell, who uses it to examine the classical cinema as a "paradigm," that is, an array of formal norms and devices that readily substitute for one another. "Both the alternatives and the limitations of the [classical] style remain clear," he writes, "if we think of the paradigm as creating *functional equivalents:* a cut-in may replace a track-in, or color may replace lighting as a way to demarcate volumes, because each device fulfills the same role. Basic principles govern not only the elements in the paradigm but also the ways in which the elements may function." David Bordwell, "An Excessively Obvious Cinema," in Bordwell, Staiger, and Thompson, *The Classical Hollywood Cinema*, 5.

31. Quoted in Leonard Maltin and Richard Bann, *The Little Rascals: The Life and Times of Our Gang* (New York: Three Rivers Press, 1992), 84.

32. On trap drumming, see Rick Altman, *Silent Film Sound* (New York: Columbia University Press, 2004), 236–40; Stephen Bottomore, "An International Survey of Sound Effects in Early Cinema," *Film History* 11, no. 4 (1999): 485–98.

33. Altman, *Silent Film Sound*, 385.

34. As early as 1911, *Moving Picture World*'s music columnist had commented on the distinction, noting: "Much liberty is allowable in comedy pictures . . . but in the straight dramatic pictures sound effects should be made to imitate as nearly as possible the real sounds which would naturally be heard in a real scene such as the picture portrays." Quoted in Altman, *Silent Film Sound*, 238.

35. For more on Jules White and Columbia's short-subjects division, see ch. 5 of this book.

36. On the use of sound effects in Columbia short comedies, see Ted Okuda and Edward Watz, *The Columbia Comedy Shorts: Two-Reel Hollywood Film Comedies, 1933–1958* (Jefferson, NC: McFarland, 1986), 42–43. By around 1939, continuity scripts at Columbia occasionally included specific instructions on Foley effects. An example would be the continuity for the Stooges' *Oily to Bed, Oily to Rise* (October 1939), which includes capitalized instructions for, e.g., a "RASPING SCREECH" when a saw runs over Curly's head, "SOUND OF BASS DRUM" when a door hits him in the rear, and an "OLD-FASHIONED HORN" and "LITTLE FRENCH HORN" when Moe, Larry, and Curly bop each other on their noses. Final draft script, February 6, 1939, 5, 8, 28, *Oily to Bed, Oily to Rise,* JWC.

37. Bruskin, *The White Brothers*, 262.

38. "Jack White Uncovers New Entertainment in Talkies."

39. For more on Sennett's burlesque melodramas, see my book *The Fun Factory: The Keystone Film Company and the Emergence of Mass Culture* (Berkeley: University of California Press, 2009), 52–63.

40. The same misapprehension—with Normand again doing the misapprehending—forms the basis of the Arbuckle-directed Keystone two-reeler *Mabel and Fatty's Married Life* (February 1915).

41. Sennett's scripting suggestions for this sequence are recorded in the story conference notes dated June 28, 1929, *The Constabule*, Production Files, MSC.

42. Dialogue quoted from script, second revision, "Iron Horses," June 26, 1929, 14, *The Constabule*, Production Files, MSC.

43. On divisions between sophisticated and popular humor, see Henry Jenkins, *What Made Pistachio Nuts? Early Sound Comedy and the Vaudeville Aesthetic* (New York: Columbia University Press, 1992), ch. 2; Lea Jacobs, *The Decline of Sentiment: American Film in the 1920s* (Berkeley: University of California Press, 2008), ch. 3.

44. The quoted phrases are from "What the Picture Did for Me," *MPH*, June 17, 1933, 49; September 9, 1933, 46.

45. Gilbert Seldes, "American Humor," in Fred J. Ringel, ed., *America as Americans See It* (New York: Literary Guild, 1932), 347–60.

46. Richard Lewis Ward, *A History of the Hal Roach Studios* (Carbondale: Southern Illinois University Press, 2005), 62–65 and 84–98.

47. "Financial Statements," *Film Daily Yearbook 1928* (New York: Film Daily, 1928), 806; Earle W. Hammons in Kennedy, ed., *The Story of the Films*, 168. Although its chief interests were in distribution, Educational had operated studio facilities for its producers since the early 1920s. The company formalized its production interests in 1927 by reorganizing itself as two companies—Educational Film Exchanges, Inc. (representing its distribution network), and Educational Pictures, Inc. (representing its production operations). Throughout this book, I have followed established film historical practice by using "Educational Pictures" to refer to the company as a whole.

48. Hammons in Kennedy, ed., *The Story of the Films*, 157. For more on this lecture series, see Beauchamp, *Joseph P. Kennedy Presents*, 93–99; Dana Polan, *Scenes of Instruction: The Beginnings of the US Study of Film* (Berkeley: University of California Press, 2008), 113–74.

49. Jack White quoted in Bruskin, *The White Brothers*, 119. When the *Film Daily Yearbook* of 1928 polled over thirty "leaders of the industry" on their predictions for the coming year, the vice president of Christie Studios, Charles Christie, used the forum to complain about precisely this issue: "Last season the terrific struggle in the two-reel comedy field was not a matter of product but a matter of bookings, and the leaving out of comedies in the bigger theaters. . . . The product is there. All that remains is getting comedies back on the screen in the houses where they have been left out." "Leaders See Prosperity for 1928," *Film Daily Yearbook 1928*, 510.

50. Advertisement for World Wide Pictures, *Film Daily Yearbook 1929* (New York: Film Daily, 1929), 87.

51. "Educational Stock Jumps," *MPH*, June 20, 1931, 12.

52. "Sennett to Direct Feature," *MPH*, May 16, 1931, 26; "Sennett to Stay with Educational Despite Rumors," *MPH*, March 21, 1931, 19. *Hypnotized* was not the only Sennett feature

released during his Educational period. Sennett had earlier directed a five-reel picture, *Midnight Daddies,* in the spring of 1929, for release under the World Wide banner. The film languished for some fifteen months before receiving only a limited release in August 1930. On Moran and Mack's careers in early sound film, see Nicholas Sammond, "As the Crow Flies: The Intermediality of Moran and Mack," *Studies in American Humor* 1, no. 2 (2015): 142–62.

53. "Named to Investigate Fox–World Wide Issue," *MPH,* September 16, 1933, 57; "World Wide Films," *MPH,* June 23, 1934, 8.

54. "Hammons Strikes at Double Bill Evil," *MPH,* December 26, 1931, 10; "Forecast for 1932," *Film Daily Yearbook 1932* (New York: Film Daily, 1932), 41. Educational also sought to combat the double bill trend by unveiling plans in late 1931 for a series of three- or four-reel *Mack Sennett Comedy Featurettes* that could be rented in place of a second feature. The plan did not come to pass, however, and the films were ultimately released as regular two-reelers. See "Shorts of Three and Four Reels Planned by the Big Producers," *MPH,* December 12, 1931, 25. See also ch. 4 on Hal Roach's similar experiments with the three-reel format during this period.

55. "Forecast for 1933," *Film Daily Yearbook 1933* (New York: Film Daily, 1933), 100.

56. Advertisements for Educational, *MPH,* April 2, 1932, 45; April 16, 1932, 47; April 30, 1932, 37; May 7, 1932, 35.

57. On small-town theaters and the family audience, see Kathryn H. Fuller-Seeley, "'What the Picture Did for Me': Small-Town Exhibitors' Strategies for Surviving the Great Depression," in Kathryn H. Fuller-Seeley, ed., *Hollywood in the Neighborhood: Historical Case Studies of Local Moviegoing* (Berkeley: University of California Press, 2008), 194.

58. The "doubleitis" ads can be found in *MPH,* July 3, 1937, 85; July 17, 1937, 107; July 31, 1937, 101; and August 14, 1937, 121.

59. See ch. 2.

60. "Hammons Hits Double Bills as Ruinous," *Motion Picture Daily,* August 11, 1932, 1, 6, quoted in Gary D. Rhodes, "'The Double Feature Evil': Efforts to Eliminate the American Dual Bill," *Film History* 23 (2011): 67; advertisement for Educational, *MPH,* July 2, 1932, 43.

61. All quotes in this paragraph are from the advertisement for Educational's 1932–1933 program, *MPH,* July 2, 1932, 43–48.

62. It is not entirely clear why Hammons failed to affiliate with a major studio distributor sooner. As early as 1927, during his lecture at Harvard, Hammons was asked precisely this question, answering: "We have been approached several times on that subject, and I had to decide no." His subsequent explanation is vague but implies that his hands were tied by theater chains with substantial interests in Educational's exchanges. "When we first produced pictures ourselves, we released through independents on a percentage basis, but we did not get the percentage we were entitled to. We finally opened our own branches. I went to the big theatre chains in the various districts. Here in this New England territory Mr. Gordon owned the biggest chain of theatres. I sold him a forty-nine per cent interest in this particular exchange, retaining a fifty-one per cent interest. In Chicago we went to Balaban and Katz and sold them a forty-nine per cent interest in that exchange." Hammons also seems to have felt that affiliation with the majors was simply not necessary for Educational's corporate health. "The business we have received from the Famous Players theatres has been about one-tenth of our gross. We are in thirteen thousand five hundred theatres. Such a combination as you speak of might be a good thing and it might not. It is really a debatable question. I am not always sure in my mind that I have made

the right decision. I hope time will prove that I have." By 1933, the decision was clearly the wrong one. See Kennedy, ed., *The Story of the Films,* 167–68.

63. "Strengthening of Fox Awaited from Financial Reorganization," *MPH,* April 8, 1933, 11.

64. "Educational Increases First-Run Bookings," *MPH,* May 6, 1933, 31.

65. See "Fewer Two-Reelers in 1935–36 Due to Duals," *MPH,* April 27, 1935, 30: "Following Paramount's decision of one week ago to abandon the two-reeler in favor of singles, until the prevailing double feature trend subsides, Columbia, Metro, Radio and Universal gave consideration to a similar solution of the double feature problem as it affects short subject sales."

66. Jack White quoted in Bruskin, *The White Brothers,* 136.

67. "Educational Seeks New Releasing Ally," *MPH,* January 22, 1938, 30.

68. James R. Shortridge, *The Middle West: Its Meaning in American Culture* (Lawrence: University Press of Kansas, 1989).

69. The phrase "those who stayed behind" is borrowed from Hal Barron, *Those Who Stayed Behind: Rural Society in Nineteenth-Century New England* (New York: Cambridge University Press, 1984).

70. Carey McWilliams, *The New Regionalism in American Literature* (Seattle: University of Washington Bookstore, 1930).

71. See Michael Denning, *The Cultural Front: The Laboring of American Culture in the Twentieth-Century* (New York: Verso, 1997), 132–36.

72. Figures cited in Gomery, *The Coming of Sound,* 93.

73. Figures cited in Fuller-Seeley, "'What the Picture Did for Me,'" 188.

74. On the history of this column, see ibid., 186–207.

75. "Broadway and/or United States," *MPH,* June 3, 1933, 19.

76. On this association of hokum, see Jacobs, *The Decline of Sentiment,* 82–83. See also my introduction.

77. Virginia Morris, "We—Want—Hokum!" *Picture Play,* April 1927, 20.

78. "Hokum Still Best, Says Fox Ad Chief," *MPH,* February 6, 1932, 25. Universal Pictures head Carl Laemmle noted this ambivalence in a curious letter to *Exhibitors Herald-World* in which he asked for readers' help in defining "hokum": "I have heard people apply the term Hokum as something good; others apply it as something bad. Because of this tremendous and radical difference of opinion, it seems to me that this industry should decide for itself what Hokum is and should give that definition to the dictionaries as a contribution from the moving picture industry. . . . Will you help me?" "What Does 'Hokum' Really Mean? Carl Laemmle Wants to Know," *EHW,* August 24, 1929, 26.

79. "Small Town Movie Fan Is Called Real Censor," *Virginian-Pilot,* December 24, 1924, 14, quoted in Terry Lindvall, "Cinema Virtue, Cinema Vice: Race, Religion, and Film Exhibition in Norfolk, Virginia, 1908–1922," in Fuller-Seeley, ed., *Hollywood in the Neighborhood,* 91.

80. Seldes, "American Humor," 360; Constance Rourke, *American Humor: A Study of the National Character* (New York: Harcourt Brace Jovanovich, 1959 [1931]), 227.

81. Harold Ellis Jones and Herbert S. Conrad, "Rural Preferences in Motion Pictures," *Journal of Social Psychology* 1, no. 3 (1930): 420–21 (emphasis added). Similarly, film historian Henry Jenkins has shown how Hollywood's "Broadway strategy" played very differently in urban and regional markets, with the ethnic-themed performances of Ziegfeld stars like Eddie Cantor meeting significant resistance from hinterland audiences. Henry Jenkins,

"'Shall We Make It for New York or for Distribution?' Eddie Cantor, *Whoopee*, and Regional Resistance to the Talkies," *Cinema Journal* 29, no. 3 (Spring 1990): 32–52. The article also appears in somewhat expanded form as chapter 6 of Jenkins's book, *What Made Pistachio Nuts?*, 153–84.

82. Peter Stanfield, *Hollywood, Westerns and the 1930s: The Lost Trail* (Exeter: University of Exeter Press, 2001).

83. On radio comedy shows with rural settings, see Tim Hollis, *Ain't That a Knee-Slapper: Rural Comedy in the Twentieth Century* (Jackson: University of Mississippi Press, 2008), 7–62.

84. "What the Picture Did for Me," *MPH*, February 4, 1933, 54.

85. "What the Picture Did for Me," *MPH*, January 26, 1933, 68.

86. For more information on the Hamilton scandal, see Anthony Balducci, *Lloyd Hamilton: Poor Boy Comedian of Silent Cinema* (Jefferson, NC: McFarland, 2009), chs. 16 and 17.

87. Information on Burke is taken from his clippings file, BRTC, and from the contracts and agreements between Mack Sennett, Inc., and John E. Burke, February 6, 1929, and December 15, 1929, Contract Files—Biography, MSC.

88. Script, no title, December 4, 1928, 2, *The Bride's Relations*, Production Files, MSC.

89. Jules White quoted in Bruskin, *The White Brothers*, 266. Clyde's screen persona was elsewhere identified as a "'rube' character" in "Mack Sennett's Latest Talking Comedy Scores," press sheet for *The Bees' Buzz*, 3.

90. The Gribbon-Clyde costarring films consisted of the following: *Whirls and Girls* (February 1929), *The Bees' Buzz* (April), *The Big Palooka* (May), *The Constabule* (August), *The Lunkhead, The Golfers* (both September), *A Hollywood Star* (October), *Clancy at the Bat, The New Halfback* (both November), *"Uppercut" O'Brien* (December), and *Sugar Plum Papa* (February 1930). After this, Gribbon left regular employment at Sennett, instead working for him on a picture-by-picture basis in such titles as *Hollywood Theme Song* (December 1930), *Dance Hall Marge* (January 1931), *The Great Pie Mystery* (October 1931), and *Hatta Marri* (July 1932). See the Harry Gribbon folder, Contract Files—Biography, MSC.

91. The designation "village boy" is taken from "The Constabule," press sheet for *The Constabule*, 1, EPS.

92. Script, "Homer's Boner," 18, *The Big Palooka*, Production Files, MSC; script draft, "Restaurant Story," July 13, 1929, 4, *The Lunkhead*, Production Files, MSC.

93. Script, "Football Story," 3, *The New Halfback*, Production Files, MSC.

94. See Charles Tepperman, "Digging the Finest Potatoes from Their Acre: Government Film Exhibition in Rural Ontario, 1917–1934," in Fuller-Seeley, ed., *Hollywood in the Neighborhood*, 130–48. On the use of rube characterizations in medicine shows, see Brooks McNamara, *Step Right Up* (Garden City, NY: Doubleday, 1976), 138.

95. "What the Picture Did for Me," *MPH*, September 28, 1935, 357.

96. Albert Blumenthal, *Small-Town Stuff* (Chicago: University of Chicago Press, 1932), 34, 398.

97. Script, 'Iron Horses,' May 1929, 7, *The Constabule*, Production Files, MSC.

98. See Kathryn H. Fuller-Seeley, "Provincial Modernity? Film Exhibition at the 1907 Jamestown Exposition," in Staiger and Hake, ed., *Convergence/Media/History*, 59–68.

99. "What the Picture Did for Me," *MPH*, July 15, 1933, 84; and April 7, 1934, 83.

100. "What the Picture Did for Me," *MPH*, August 19, 1933, 55; "What the Picture Did for Me," *MPH*, February 3, 1934, 71; and "What the Picture Did for Me," *MPH*, n.d., n.p., Columbia—Clippings, JWC. A comparison with Will Rogers would seem to suggest itself, since both Clyde and Rogers offered variants on the "wise rube" tradition during this period. Yet what distinguished Rogers's comedy was, of course, the political mileage he drew from that tradition, using the fool's simplicity and fair-mindedness as a vantage point to evoke an inclusive left-wing populism. See, for instance, Lary May, *The Big Tomorrow: Hollywood and the Politics of the American Way* (Chicago: University of Chicago Press, 2000), ch. 1. Only once did Clyde's comedies approach overt political critique, and from a very different ideological perspective, in the later Columbia short, *Share the Wealth* (March 1936). See ch. 5, n. 86.

101. Clyde's "old man" persona thus appeared on movie screens consistently for a twenty-seven-year period, from 1929 to 1956, just besting Chaplin's twenty-six-year stretch as the little tramp (from Keystone's *Mabel's Strange Predicament* in 1914 to *The Great Dictator* in 1940).

102. At a time when Educational's roster was becoming particularly depleted of name comic talent—the increasingly alcoholic and financially troubled Lloyd Hamilton having been let go in 1931, while Sennett departed for Paramount the following year—the *Star Personality* series was announced late in 1933 with the intent of signing up "the big names of radio, stage, and screen . . . to reinforce Educational's established favorites." Advertisement for Educational, *MPH*, November 11, 1933, 37. The series was also evidently designed to make the best of Educational's partial retrenchment to the East Coast by using the opportunity to sign up New York–based talent. (Early stars in the series, for instance, included the CBS radio comedy double-act "Easy Aces"—husband and wife team Goodman and Jane Ace—and Broadway comics Joe Cook and Milton Berle.)

103. The eleven films are, in order of release, *The Gold Ghost* (March 1934), *Palooka from Paducah* (January 1935), *One Run Elmer* (February), *Hayseed Romance* (March), *The E-Flat Man* (August), *The Timid Young Man* (October), *Three on a Limb* (January 1936), *Grand Slam Opera* (February), *Blue Blazes* (August), *Ditto* (February 1937), and *Love Nest on Wheels* (March).

104. "What the Picture Did for Me," *MPH*, January 19, 1935, 87; February 20, 1937, 76; May 1, 1937, 74.

105. "What the Picture Did for Me," *MPH*, July 13, 1935, 84; June 8, 1935, 109; and February 2, 1935, 91.

106. Review of *The Screen Test*, *MPH*, December 19, 1936, 59; review of *Happy Heels*, *MPH*, August 22, 1936, 53.

107. "J. C. Jenkins—His Colyum," *EHW*, Nov. 2, 1929, 71.

108. Another example—albeit a far less successful one—would be Hal Roach's proposed series of seven shorts starring southern humorist Irvin S. Cobb in 1934–1935, to be released as part of Roach's *All-Star* comedy line. A kind of Kentucky-fried version of W. C. Fields-style domestic comedy, the series proved so unpopular with exhibitors that it was ended after only four films, one of which was not even released to theaters.

109. Pierre Bourdieu, *The Rules of Art: Genesis and Structure of the Literary Field*, trans. Susan Emanuel (Stanford, CA: Stanford University Press, 1995), 254.

110. Ibid., 255.

CHAPTER 4. "I WANT MUSIC EVERYWHERE"

1. Annual Report, August 31, 1928, Files—1930s, HRC.

2. "Hal Roach Getting Equipment to Set Comedies in Sound," *EHMPW,* June 9, 1928, 67; "Hal Roach Allies with Victor for Sound Production," *EHMPW,* October 13, 1928, 32. Irrespective of Loew's-MGM's guidance, however, Roach seemingly had faith early on that sound was the future. Even prior to the above-quoted report, the company had invested in sound-on-disc musical tracks with synchronized sound effects for its shorts, beginning with the April 1928 *Our Gang* release, *Barnum and Ringling, Inc.*

3. On Raguse, see "Victor Expert at Hal Roach Studio for Sound Stages," *EHMPW,* December 1, 1928, 36; Randy Skretvedt, *Laurel and Hardy: The Magic behind the Movies* (Beverly Hills, CA: Moonstone Press, 1987), 157–59.

4. See, for example, Michele Hilmes, *Radio Voices: American Broadcasting, 1922–1934* (Minneapolis: University of Minnesota Press, 1997); Ross Melnick, *American Showman: Samuel "Roxy" Rothafel and the Birth of the Entertainment Industry* (New York: Columbia University Press, 2012); Katherine Spring, *Saying It with Songs: Popular Music and the Coming of Sound to Hollywood Cinema* (New York: Oxford University Press, 2013); Steve J. Wurtzler, *Electric Sounds: Technological Change and the Rise of Corporate Mass Media* (New York: Columbia University Press, 2007).

5. Melnick, *American Showman,* 307.

6. See James P. Kraft, "Musicians in Hollywood: Work and Technological Change in Entertainment Industries, 1926–1940," *Technology and Culture* 35, no. 2 (April 1994): 289–314.

7. Ann Douglas, *Terrible Honesty: Mongrel Manhattan in the 1920s* (New York: Farrar, Strauss & Giroux, 1995), 376.

8. Susan Stewart, *On Longing: Narratives of the Miniature, the Gigantic, the Souvenir, the Collection* (Durham, NC: Duke University Press, 1993), 65.

9. Doane to Roach, November 27, 1928, Files—1930s, HRC.

10. Ibid.

11. "35 of 40 Hal Roach Comedies to Have Synchronization," *MPN,* September 1, 1928, 725, quoted in Richard Lewis Ward, *A History of the Hal Roach Studios* (Carbondale: Southern Illinois University Press, 2005), 73.

12. Lynch to Roach, February 13, 1930, Files—1930s, HRC.

13. Hal Roach Comedy Team to Do Victor Song Record," *EHW,* June 1, 1929, 45; "W. W. Clark and Hal Roach Discuss All-Musical Films," *EHW,* June 22, 1929, 142. Although tests of "Honey" were recorded on April 27, 1929 (with Leroy Shield on piano), Todd was eventually removed from the production of *Dad's Day* and no phonograph was made. Studio logs, April 29, 1929, www.leroyshield.com (accessed September 18, 2012). The Hal Roach Studios did, however, successfully launch a series of musical shorts a few years later, for the 1933–1934 season. The "Schmaltz brothers" films are discussed later in this chapter.

14. Lynch to Roach, October 29, 1930, Files—1930s, HRC.

15. Roach to Lynch, November 5, 1930, and Lynch to Roach, May 12, 1931, Files—1930s, HRC.

16. Donald Crafton, *The Talkies: American Cinema's Transition to Sound, 1926–1931* (New York: Charles Scribner's Sons, 1997), 315–16.

17. A case in point is offered by Mack Sennett, whose unsuccessful attempts to capitalize on the song tie-in vogue for the Educational short *Radio Kisses* (May 1930) required cold-

calling music publishers to gauge their interest in his self-penned tunes (cowritten with comedian and gag writer Harry McCoy). Sennett received three rejection letters, one of which criticized the songs for "too much repetition." Herbert E. Marks, Edward B. Marks Music Co., to Jed Buell, May 13, 1930; J.M. Davis, Triangle Music Publishing Co., to Jed Buell, May 13, 1930; Chas. Lang, Bibo-Lang Music Publishers, to Jed Buell, May 15, 1930, *Radio Kisses,* Production Files, MSC. It is tempting to hypothesize that the rejections fueled Sennett's brilliant takedown of the theme-song trend in the Arthur Ripley–scripted *Hollywood Theme Song* (December 1930): there, a small-town hero (Harry Gribbon) constantly bursts into songs that comment on his actions, accompanied by three musicians who lug their instruments after him wherever he goes. Sennett subsequently hit the musical jackpot the following year when he signed two performers from Gus Arnheim's Cocoanut Grove orchestra in 1931: Irish tenor Don Novis and then up-and-comer Bing Crosby. The resulting Crosby shorts, in particular, typify what historian Katherine Spring calls the "star-song attractions" of the conversion era, when films often served as vehicles for stars to perform their trademark songs: Crosby's debut Sennett short, *I Surrender Dear* (September 1931) thus includes his song "I Surrender Dear," while his second, *One More Chance* (November 1931), prominently features "Just One More Chance." See Spring, *Saying It with Songs,* 24–29 and ch. 3.

18. Subsequent information on Hatley's and Shield's careers is synthesized from the following sources: Piet Schreuders, liner notes from the Beau Hunks CDs, *The Beau Hunks Play the Original Laurel & Hardy Music* (Movies Select Video, 1992) and *On to the Show!* (BASTA, 2007 [1995]); Jim Shadduck, "The Ku-Ku Song Man!," *Pratfall* 7 (1972), n.p., http://archive.is/p6pcZ (accessed February 21, 2014); and Warren M. Sherk, "The Dance of the Cuckoos: Music in the Films of Lauren and Hardy," *Cue Sheet* 14, no. 4 (October 1998): 28–34.

19. Laurel quoted in Sherk, "Dance of the Cuckoos," 29–30.

20. Sound engineer Elmer Raguse himself implemented a kind of jerry-built version of wall-to-wall underscoring early in 1930 by mixing preexisting commercial records into the soundtracks of the studio's releases, albeit with no concern for synchronization: a record would be played twice, changed for a new one, which would in turn be played twice, and so forth, until the end of the film. My thanks to Richard W. Bann for clarity on these early scoring initiatives.

21. Cutting continuity, undated, *Another Fine Mess,* Production Files, George Stevens Collection, AMPAS.

22. Shield to Ginsberg, April 11, 1935, Leroy Shield correspondence, 1932–1937, HRC.

23. Schreuders, liner notes from *The Beau Hunks Play the Original Laurel & Hardy Music,* 4. The specifics of the mixing technology used at Roach are unclear: Piet Schreuders comments that "It is believed that [Shield's tunes] were all recorded directly onto film, copies of which were then stored in a kind of jukebox, a machine holding up to fifty or sixty film loops at any one time. A sound operator could bring up a tune at the touch of a button." Schreuders, liner notes from *The Beau Hunks Play the Original Laurel & Hardy Music,* vol. 2 (Movies Select Video, 1993), 4.

24. In this respect, underscoring practice at Roach mirrored early experiments with incidental music at other studios around this time. As Michael Slowik notes, music was conceived "not as an object to be molded and reworked to fit with specific actions or lines of dialogue but rather as an autonomous entity that should coincide in a segment-by-segment manner." Michael Slowik, *After the Silents: Hollywood Film Music in the Early Sound Era, 1926–1934* (New York: Columbia University Press, 2014), 94.

25. Shield to R.S. Peer, September 11, 1930, www.leroyshield.com (accessed January 29, 2014).

26. Henry Ginsberg to Leroy Shield, February 18, 1932, Leroy Shield correspondence, 1932–1937, HRC. Historian Warren Sherk hypothesizes that Shield's asking price of $500 per week may have nixed the deal. Sherk, "Dance of the Cuckoos," 31.

27. Henry Ginsberg to Leroy Shield, October 15, 1935, Leroy Shield correspondence, 1932–1937, HRC.

28. Hal Roach to Leroy Shield, February 18, 1936, Leroy Shield correspondence, 1932–1937, HRC. This letter was in response to a telegram dated two days earlier in which Shield explained: "IN DECEMBER HENRY GINSBERG WROTE ME THAT YOU WERE GOING TO ESTABLISH A MUSIC DEPARTMENT I ANSWERED SAYING I WOULD LIKE TO COME OUT STOP HAVE HEARD NOTHING FURTHER BUT WOULD LIKE TO WORK FOR YOU VERY MUCH." Leroy Shield to Hal Roach, telegram, February 16, 1936, Leroy Shield correspondence, 1932–1937, HRC.

29. Matt O'Brien to Leroy Shield, April 23, 1936, Leroy Shield correspondence, 1932–1937, HRC.

30. Spring, *Saying It with Songs*, ch. 5.

31. Ibid., 123.

32. On *The Squall*, see Slowik, *After the Silents*, ch. 3. Slowik also notes that early part-talkie films often featured extensive underscoring of dialogue scenes.

33. Steiner's early scores date to the 1931–33 period, during his time at RKO, in films like *Symphony of Six Million* (1932), *Bird of Paradise* (1932), and *King Kong* (1933).

34. This understanding of studio-era compositional practice as modeled on classical musical principles is well exemplified by the work of Kathryn Kalinak. "The medium of the classical Hollywood film score was largely symphonic," Kalinak writes, "its idiom romantic; and its formal unity typically derived from the principle of the leitmotif." Kathryn Kalinak, *Settling the Score: Music and the Hollywood Film* (Madison: University of Wisconsin Press, 1992), 79. See also Caryl Flinn, *Strains of Utopia: Gender, Nostalgia, and Hollywood Film Music* (Princeton, NJ: Princeton University Press, 1992). The applicability of a classical model to studio-era scoring has, however, been critiqued by Jennifer Fleeger, *Sounding American: Hollywood, Opera, and Jazz* (New York: Oxford University Press, 2014), ch. 4; and Scott D. Paulin, "Richard Wagner and the Fantasy of Cinematic Universality: The Idea of the *Gesamtkunstwerk* in the History and Theory of Film Music," in James Buhler, Caryl Flinn, and David Neumeyer, eds., *Music and Cinema* (Hanover, NH: Wesleyan University Press, 2000), 58–84.

35. See ch. 3.

36. "A Showman Discusses the Short-Comings of the Short Feature," *MPH*, April 23, 1932, 49.

37. Michel Chion, *Audio-Vision: Sound on Screen*, trans. Claudia Gorbman (New York: Columbia University Press, 1994), 13.

38. Hatley quoted in Skretvedt, *Laurel and Hardy*, 197.

39. Hatley quoted in ibid.; also in Jim Shadduck, "The Ku-Ku Song Man!," n.p.

40. Review of *Berth Marks*, *MPN*, October 5, 1929, 1264. The pacing of action sequences lacking dialogue was a widely acknowledged problem during the conversion period, drawing specific commentary from one of the earliest advocates of background scoring, Welford C. Beaton, editor of *Film Spectator*. Prior to the popularization of underscoring, Beaton wrote in 1932, an action sequence without dialogue unfolded "like something that had died and

was following its own hearse." Welford Beaton, *Know Your Movies: The Theory and Practice of Motion Picture Production* (Hollywood, CA: Howard Hill, 1932), 85.

41. Fleeger, *Sounding American,* 136.

42. A reference to one of Bourdieu's more trenchant observations in his 1992 study *The Rules of Art* can perhaps sharpen our understanding of the factors at play here. As he suggests there, the autonomy of any field of production—let us say, slapstick cinema—ensures that, in the works produced in it, any contextualizing historical change—be it social, political, or, in the case of sound, technological/industrial—will always be filtered through the configuration of values and norms that already exist within that field. "The direction of [any] change depends on the state of the system of possibilities (conceptual, stylistic, etc.) inherited from history." Bourdieu, *The Rules of Art: Genesis and Structure of the Literary Field,* trans. Susan Emanuel (Stanford, CA: Stanford University Press, 1995), 205. One cannot, that is, point simply to the convergence of film and music industries as a "cause" of which certain scoring practices were isolatable "effects" without also taking into account the "system of possibilities"—established norms of craft, etc.—within the various filmmaking fields that in turn shaped the direction of musical innovation. If the slapstick short was the first field of live-action filmmaking to witness the institutionalization of musical underscoring, then this was because sound's advent had created pressing problems relating to comedians' performative craft to which music could lend resolution.

43. Hal Roach to Felix Feist, October 6, 1927, 1926–1928, HRC.

44. Financial statement, April 25, 1932, in Hal Roach Studios, Inc., Company Minutes, September 21, 1931 to November 25, 1932, 67, HRC.

45. The statistics for 1933–1934 and 1934–1935 are from Ward, *A History of the Hal Roach Studios,* 85.

46. Ward, *A History of the Hal Roach Studios,* 80–81.

47. Roach quoted in Skretvedt, *Laurel and Hardy,* 202.

48. "Famed Prisons Duplicated in 'Pardon Us,'" *Pardon Us* press sheet, 1, BRTC.

49. Hal Roach Studios, Inc., Company Minutes, June 25, 1927 to August 25, 1931, 193, HRC.

50. Box-office gross cited in Ward, *A History of the Hal Roach Studios,* 211.

51. Contracts between MGM Distributing Corp. and Hal Roach Studios, Inc., dated January 5, 1932, December 28, 1932, May 15, 1933, and July 20, 1934, Hal Roach Studios/MGM Contracts, Hollywood Museum Collection, USC.

52. Richard M. Roberts, *Past Humor, Present Laughter: Musings on the Comedy Film Industry, 1910–1945,* vol. 1, *Hal Roach,* co-researched with Robert Farr and Joe Moore (Phoenix, AZ: Practical Press, 2013), 189.

53. Ward, *A History of the Hal Roach Studios,* 82–83. The other two features were *The Outsider* (1931), a British film, and *Flirting with Fate* (1938), which was produced by David Loew, son of Loew's founder Marcus Loew.

54. The challenge of feature-length slapstick has been a frequent area of investigation for film scholars, albeit typically with a focus on the silent era. See, for instance, my book *The Fun Factory: The Keystone Film Company and the Emergence of Mass Culture* (Berkeley: University of California Press, 2009), ch. 3, as well as Steve Neale and Frank Krutnik, *Popular Film and Television Comedy* (London: Routledge, 1990), ch. 6. Such discussions have tended to center on the tension between gags, understood as forms of narrative disruption or delay, and the demands of narrative clarity and progress. See also, on this tension, Donald Crafton, "Pie and Chase: Gag, Spectacle and Narrative in Slapstick

Comedy," in Kristine Karnick and Henry Jenkins, eds., *Classical Hollywood Comedy* (New York: Routledge, 1995), 106–19; and Peter Krämer, "Derailing the Honeymoon Express: Comicality and Narrative Closure in Buster Keaton's *The Blacksmith*," *Velvet Light Trap* 23 (Spring 1989): 101–16.

55. "Jailbirds," unidentified clipping (presumably UK); and "Jailbirds," unidentified clipping (presumably UK), *Pardon Us* clippings file, BRTC.

56. Marguerite Tazelaar, "Pack Up Your Troubles," *NYHT,* October 1, 1932, n.p., *Pack Up Your Troubles* clippings file, BRTC.

57. "The Empire," *Times* (London), October 17, 1932, n.p., *Pack Up Your Troubles* clippings file, BRTC.

58. Added Scenes (6 pp.) for production F-2, Laurel and Hardy—Scripts and Production Material, Hal Roach Studios Collection, AMPAS.

59. Felix Feist to Henry Ginsberg, August 25, 1932, Files—1930s, HRC.

60. Laurel and Hardy had already cameoed in the nonnarrative, revue-style MGM feature *The Hollywood Revue of 1929* (1929), in which they performed inept conjuring tricks. The team's appearance in *The Rogue Song* seems to have been the result of a last-minute decision, made midway into the film's two-month shooting schedule. According to production reports, shooting on the film lasted from August 22 to October 18, 1929; Laurel and Hardy were brought in for a test on September 21 and shot their scenes in early October, most filmed separately from the main cast. Despite this, the duo received third-place billing in credits and publicity for the film, following Tibbett and love interest Catherine Dale Owen. See Daily Production Reports and Main Title Billing, *The Rogue Song,* Box 109, MGM Collection, USC.

61. "Sound Pictures Alter Musical Taste, in Opinion of Lawrence Tibbett," souvenir programme, *The Rogue Song,* 12, BRTC. On ideologies of musical uplift and sound technologies, see Wurtzler, *Electric Sounds,* ch. 4.

62. "The Rogue Song," unidentified clipping, and Edwin Schallert, unidentified clipping, Audrey Chamberlin scrapbooks, #28 of 46, 98, 100–101, AMPAS.

63. Crafton, *The Talkies,* 320. It is further possible that MGM also had in mind RKO's 1929 musical *Rio Rita,* which also features a subplot with two comedians. The RKO film was based on the 1927 Broadway musical of the same name, which had originally united Bert Wheeler and Robert Woolsey in secondary roles that they repeated for the film version. (They were in fact the only principals from the original play to also appear in the movie.) Based on the success of that film—RKO's biggest box-office hit until *King Kong* (1933)—Wheeler and Woolsey went on to a successful career as comic stars at RKO until Woolsey's death in 1937.

64. Jacques Attali, *Noise: The Political Economy of Music,* trans. Brian Massumi (Minneapolis: University of Minnesota Press, 1985), 122.

65. Ibid., 6, 19 (emphasis added).

66. Ibid., 59–62, 109.

67. Krin Gabbard, *Jammin' at the Margins: Jazz and the American Cinema* (Chicago: University of Chicago Press, 1996), 8–19; William T. Schultz, "Jazz," *Nation,* October 25, 1922, 438.

68. Douglas, *Terrible Honesty,* 355

69. Ibid., 358.

70. On popular music and ethnic assimilation, see, for instance, Douglas, *Terrible Honesty,* ch. 9; David R. Roediger, *Working toward Whiteness: How America's Immigrants Became White* (New York: Basic Books, 2005), ch. 4; Michael Rogin, *Blackface, White Noise:*

Jewish Immigrants in the Hollywood Melting Pot (Berkeley: University of California Press, 1996), ch. 5.

71. Rick Altman, *The American Film Musical* (Bloomington: Indiana University Press, 1987), 138–39. A very different reading is provided by Raymond Knapp, who places the operetta within his broader interpretation of American musical films as vehicles for "performing" an individualized conception of identity. For Knapp, the operetta does not *conjoin* the romantic narrative to the political question of governance, as Altman posits, so much as it *prioritizes* the former over the latter. "Operetta as a type," he writes, typically "tak[es] political situations and reveal[s] that, after all, it is indeed 'really' the personal relationships that count." Knapp, *The American Musical and the Performance of Personal Identity* (Princeton, NJ: Princeton University Press, 2006), 11.

72. As Altman writes, "Government is simply a question of proper marriage, and, conversely, marriage is but a question of proper government. In order to develop this plot literally, nearly every operetta tradition has recourse to the creation of small, imaginary kingdoms run—as in former times and distant states—according to a personal type of reign. The importance of the government/love metaphor and plot connection explains the regularity with which every operetta chooses its characters from the upper aristocracy or royalty. . . . Gilbert and Sullivan give us Japan and the Mikado, Offenbach a series of closed principalities from Gerolstein to Hades, Lehár contributes Marshovia, Luxembourg, and the Orient, and every subsequent Viennese composer follows his lead." Altman, *The American Film Musical*, 149.

73. The historic development of operetta in fact confirms the clown's narrative marginalization. By the time of the Viennese operetta vogue of the early twentieth century, the revue elements that had been so much a part of the older operetta tradition were thoroughly out of place in stories focused on romance and governance, such that Franz Lehár would typically confine comic spectacle to a handful of duets involving the clown and soubrette, usually just three per operetta. Richard Traubner, *Operetta: A Theatrical History* (New York: Routledge, 2003), 231.

74. The classic study of melodrama and bourgeois culture is, of course, Peter Brooks, *The Melodramatic Imagination: Balzac, Henry James, Melodrama, and the Mode of Excess* (New York: Columbia University Press, 1985).

75. Thornton Delehanty, "The New Film," *New York Evening Herald*, June 12, 1933, n.p.; John S. Cohen, Jr., "The Talking Pictures," *New York Sun*, June 17, 1933, n.p.; and John S. Cohen, Jr., "The New Talkie," *New York Sun*, June 18, 1933, n.p., all from *The Devil's Brother* clipping file, BRTC.

76. Marguerite Tazelaar, "On the Screen," *NYHT*, June 10, 1933, n.p., *The Devil's Brother* clippings file, BRTC.

77. "Mop and Broom Handles at Premium Because of this Game," *Babes in Toyland* pressbook, 6, BRTC.

78. Catherine Clément, "Through Voices, History," in Mary Ann Smart, ed., *Siren Songs: Representations of Gender and Sexuality in Opera* (Princeton, NJ: Princeton University Press, 200), 19–20 (emphasis in original).

79. Dialogue Cutting Continuity, *The Devil's Brother*, April 21, 1933, 9, MGM Scripts Collection, AMPAS.

80. This comic device of mismatching voices must have tickled Roach's filmmakers, who decided to repeat it for the duo's performance of "The Trail of the Lonesome Pine" in their 1937 feature *Way Out West*.

81. Mikhail Bakhtin, "Forms of Time and of the Chronotope in the Novel: Notes toward a Historical Poetics," in *The Dialogic Imagination: Four Essays* (Austin: University of Texas Press, 1981), 159.

82. "The Bohemian Girl," *Variety,* February 19, 1936, n.p., *The Bohemian Girl* clippings file, BRTC.

83. "Bonnie Scotland," *Variety,* August 28, 1935, n.p., and "Bonnie Scotland," *Liberty Magazine,* September 7, 1935, n.p., from *Bonnie Scotland* clippings file, BRTC.

84. Janney quoted in Skretvedt, *Laurel and Hardy,* 304.

85. It appears that Paramount, too, had some interest in adapting the Herbert original, at least to judge from existing studio files from the early sound period that include an undated synopsis of the operetta's plot. Synopsis by Marion Valentine, "Paramount Synopses, 1929–1938," Howard Estabrook papers, AMPAS.

86. See MGM Distributing Corp. to Hal Roach Studios, Inc., November 29, 1933, Hal Roach Studios/MGM Contracts, 1930–1933, Hollywood Museum Collection, USC; Ward, *A History of the Hal Roach Studios,* 211.

87. "Studio Transformed into Fairyland for Laurel-Hardy Feature," *Babes in Toyland* pressbook, 3, BRTC.

88. Regina Crewe, "'Babes in Toyland' Thrilling Tale for Children at Astor," *New York American,* December 14, 1934, n.p., *Babes in Toyland* clippings file, BRTC.

89. "Babes in Toyland," *Variety,* December 17, 1934, n.p., *Babes in Toyland* clippings file, BRTC.

90. "Babes in Toyland," *Time,* December 10, 1934, n.p., *Babes in Toyland* clippings file, BRTC.

91. "Child Patients Hear Talkie in Bellevue," *NYT,* December 19, 1934, 2.

92. Susan Stewart, *On Longing: Narratives of the Miniature, the Gigantic, the Souvenir, the Collection* (Durham, NC: Duke University Press, 1993), 65.

93. "Studio Transformed into Fairyland," 3. The Toyland set was indeed elaborate, requiring two joined sound stages to house a structure 250 feet wide by five hundred in length. "World's Largest Soundstage Houses Colorful Street in Wonderland," *Babes in Toyland* pressbook, 5, BRTC.

94. "New Laurel and Hardy Feature-Length Picture Has Spectacular Scenes," *Babes in Toyland* pressbook, 3, BRTC.

95. "Babes in Toyland," *Variety,* December 17, 1934, n.p., *Babes in Toyland* clippings file, BRTC.

96. See Jack Zipes, *Fairy Tales and the Art of Subversion: The Classical Genre for Children and the Process of Civilization* (New York: Routledge, 1983), ch. 5.

97. A useful comparison of MacDonough's two librettos is provided by Larry Moore in his New York Public Library blog post, "Musical of the Month: 'Babes in Toyland,'" www.nypl.org/blog/2012/01/24/musical-month-babes-toyland (accessed July 15, 2014).

98. Roach quoted in Skretvedt, *Laurel and Hardy,* 335.

99. Comedian and director Larry Semon's *Wizard of Oz* (1925), with Semon as the Scarecrow, probably deserves mention as an earlier example. But the film was far less successful, and its distributor—Chadwick Pictures—was bankrupt before the film's initial run was complete.

100. Budget figures from Ward, *A History of the Hal Roach Studios,* 211.

101. In New York City, it continues to screen for these holidays on the station WPIX (as of 2015).

102. See Jacky Bratton and Ann Featherstone, *The Victorian Clown* (Cambridge: Cambridge University Press, 2006), 9–10.

103. Michel Foucault, "Of Other Spaces," *Diacritics* 16, no. 1 (Spring 1986): 23–24.

104. "At the Rialto," *Brooklyn Daily Eagle*, November 12, 1936, n.p., and Frank S. Nugent, "Our Relations," *NYT*, November 11, 1936, n.p., *Our Relations* clippings file, BRTC; untitled review, *New Yorker*, September 19, 1938, n.p., and "Block-Heads," *Variety*, August, 31, 1938, n.p., *Block-Heads* clippings file, BRTC.

CHAPTER 5. "FROM THE ARCHIVES OF KEYSTONE MEMORY"

1. James Agee, "Comedy's Greatest Era," in *Agee on Film: Criticism and Comment on the Movies* (New York: Modern Library, 2000), 105–6.

2. The quoted characterization is from a *Moving Picture World* review of Keystone's *Riley and Schultz* (1912), "Comments on the Films," October 12, 1912, 144.

3. Agee, "Comedy's Greatest Era," 394.

4. "Janecky Put Over Big Tie-Up When Selling a Comedy," *MPH*, April 23, 1932, 72.

5. The reissue market for short silent comedies opened up in the early 1930s when the Van Beuren Corporation began releasing Chaplin's Mutual shorts (1916–1917) through RKO. Over the decade, the market built to a point at which industry observers began to speak of a definite "trend," with new companies formed precisely for this purpose. See, for instance, "Trend toward Revival of Old Slapsticks," *MPH*, June 1, 1940, 51.

6. "Keystone Cops Return," *MPH*, July 25, 1931, 24; Vitaphone advertisement, *MPH*, September 14, 1935, 4–5. Finlayson's and Cook's careers with Sennett both postdate the Keystone period (1912–1917). Although the famous cops had been largely retired since the Keystone days, they had been occasionally revived in Sennett films up to the close of the silent era in two-reelers like *Smith's Picnic* (December 1926) and *Love in a Police Station* (December 1927).

7. Henry Ginsberg to Morton Spring, MGM Dist. Corp., NY, April 13, 1932, Correspondence Files 1930s, HRC; advertisement for MGM short subjects, *MPH*, July 9, 1932, 23 (emphasis in original).

8. On undercranking in silent comedy, see also ch. 3. Although the initial offerings in the Taxi Boy series indeed featured the promised spectacle of elaborately crashing taxi-cabs (in particular the first, *What Price Taxi* [August 1932]), budget cutbacks at the Roach Studios eventually necessitated a smaller scale, with Ben Blue and Billy Gilbert doing what sometimes amounted to a thinly veiled Laurel and Hardy impersonation. The series was not renewed for a second season.

9. "The Newest Pictures," *MPH*, June 25, 1932, 30.

10. Edward Bernds, *Mr. Bernds Goes to Hollywood: My Early Life and Career in Sound Recording at Columbia with Frank Capra and Others* (Lanham, MD: Scarecrow Press, 1999), 236.

11. Mary Douglas, "The Social Control of Cognition: Some Factors in Joke Perception," *Man* 3, no. 3 (1968): 366 (emphasis added).

12. White quoted in David N. Bruskin, *The White Brothers: Jack, Jules, & Sam White* (Metuchen, NJ: Scarecrow Press, 1990), 163.

13. See Christine Sprengler, *Screening Nostalgia: Populuxe Props and Technicolor Aesthetics in Contemporary American Film* (New York: Berghahn Books, 2008), ch. 1. I am indebted to Sprengler's superb analysis for some of the citations in this paragraph.

14. Fred Davis, *Yearning for Yesterday: A Sociology of Nostalgia* (New York: Free Press, 1979), 4–5n8. Jameson's examinations of nostalgia in the context of late capitalism have been widely published; the best known is his essay "Postmodernism, or, The Cultural Logic of Late Capitalism," *New Left Review* 146 (1984): 53–92.

15. Quotes taken from Sprengler, *Screening Nostalgia,* 23.

16. Sprengler, *Screening Nostalgia,* 26.

17. P. W. Wilson, "Victorian Days That Beckon Us," *NYT,* April 10, 1933, SM10, quoted in Sprengler, *Screening Nostalgia,* 27.

18. The series, which ran only for the 1932–1933 season, consisted of what *Motion Picture Herald* described as "old newsreel shots [i.e., actualities] of important events back in the gay old '90s," with voiceover commentary supplied by Lew Lehr. James Cunningham, "Asides & Interludes," *MPH,* October 8, 1932, 33.

19. Bosley Crowther, "Annual Message: Mr. Hays's Report Speaks a Word for the 'Therapeutic Value' of Entertainment," *NYT,* April 6, 1941, X5.

20. Film scholar Haidee Wasson describes the language with which MoMA's first circulating programs were greeted: "Films were described as being 'primitive,' 'archaic,' and 'rare' and as 'lost treasures,' 'relics,' 'antiques,' and 'ancient thrillers.' Films were 'unearthed,' 'resurrected,' 'reborn,' and 'embalmed.' The film 'veil' had been lifted. The Film Library became an 'asylum for film' and a 'sanctuary against time.' Only forty years after the first projected films, the cinema had acquired the sense of wonder and discovery usually reserved for objects of lost civilizations and faraway cultures." Haidee Wasson, *Museum Movies: The Museum of Modern Art and the Birth of Art Cinema* (Berkeley: University of California Press, 2005), 172.

21. "Remaking of 29 Hits Points to New Trend," *MPH,* October 5, 1935, 13–14.

22. On the "Great Hokum Mystery" series, see also ch. 3. Consisting of clips excerpted from the private collection of the late Abram Stone, the "Flicker Frolics" films seem to have been released in at least two batches, in 1936 and 1942, the latter for nontheatrical exhibition.

23. "Movie of the Week: Hollywood Cavalcade," *Life,* October 9, 1939, 63–69.

24. "The Hollywood Scene," *MPH,* May 20, 1939, 47; "Picture Pioneers," *MPH,* June 3, 1939, 9.

25. "The Hollywood Scene," 47.

26. Davis, *Yearning for Yesterday,* 49.

27. I am here paraphrasing Pierre Bourdieu's discussion of the "social aging" of cultural forms: "Fossilized artists are in some way old twice over, by the age of their art and their schemas of production but also by the whole lifestyle of which the style of their works is one dimension." Pierre Bourdieu, *The Rules of Art: Genesis and Structure of the Literary Field,* trans. Susan Emanuel (Stanford, CA: Stanford University Press, 1996), 150.

28. "Columbia Corners Comics for Shorts," n.s., n.d., n.p., in "Columbia—Clippings," 9, JWC.

29. Gilbert Seldes, *The 7 Lively Arts* (New York: Sagamore Press, 1957 [1924]), 31.

30. On this aspect of the Stooges shorts, see, for instance, Peter Brunette, "The Three Stooges and the (Anti-)Narrative of Violence: De(con)structive Comedy," in Andrew Horton, ed., *Comedy/Cinema/Theory* (Berkeley: University of California Press, 1991), 174–87.

31. "Comments on the Films," *Moving Picture World,* June 13, 1914, 1541.

32. White quoted in Bruskin, *The White Brothers,* 201, 231.

33. White quoted in Ted Okuda and Edward Watz, *The Columbia Comedy Shorts: Two-Reel Hollywood Film Comedies, 1933–1958* (Jefferson, NC: McFarland, 1986), 25.

34. On the White brothers' childhood in Edendale, see Bruskin, *The White Brothers*, Book I.

35. Richard Schmidt, "Interview with Jules White," *Journal of Popular Film* 6, no. 1 (1977): 45. The *Dogville* comedies were not the only all-animal parodies from this period: Tiffany had a *Chimp Family* series (1930–1932), including western spoofs like *The Little Covered Wagon* (September 1930) and *Cinnamon* (October 1931).

36. "Columbia to Make Shorts," *MPH*, July 23, 1932, 93. Production stats derived from "Lists—Costs" and "Lists—1934–1958," JWC. Budgets for short subjects at Columbia remained stable at around $15,000 for most of the 1930s, before creeping up to an average of just over $20,000 by the mid-1940s.

37. White quoted in Bruskin, *The White Brothers*, 211.

38. "Columbia Abandons Short Production in Hollywood," *Variety*, August 5, 1932, n.p., in "Columbia—Clippings," 1, JWC; White quoted in Bruskin, *The White Brothers*, 212. Smith and Dale would eventually appear in two Columbia shorts later in the decade, *A Nag in the Bag* (November 1938) and *Mutiny on the Body* (February 1939).

39. Production dates from "1933–1934 Shorts Program," Columbia—Lists, JWC. On strategies of variety appeal, see ch. 2.

40. "Meyers [sic] Produces Comedies," *MPH*, October 14, 1933, 26. The Catlett films were perhaps the most sporadic of Columbia's series, with just six entries released between 1934 and 1940. Two of the films—*Get Along Little Hubby* (June 1934) and *Static in the Attic* (September 1939)—exploit the same basic domestic situation as the Edgar Kennedy shorts: husband, wife, and annoying in-laws.

41. Review of *Ants in the Pantry*, *MPH*, n.d., n.p., in "Columbia—Clippings," 2, JWC; review of *New News*, *MPH*, n.d., n.p., in "Columbia—Clippings," 10, JWC.

42. "What the Picture Did for Me," *MPH*, n.d., n.p., in "Columbia—Clippings," 4, JWC.

43. Okuda and Watz, *The Columbia Comedy Shorts*, 139–40.

44. See ch. 1.

45. On Healy suing for use of the name "Stooge," see Steve Cox and Jim Terry, *One Fine Stooge: Larry Fine's Frizzy Life in Pictures* (Nashville, TN: Cumberland House, 2006), 37. The circumstances surrounding Healy's death have drawn much speculation. What is known is that he died on December 21, two days following a drunken fistfight at the Trocadero on Sunset Boulevard, where he had been celebrating news of his wife's pregnancy. (His assailant has been reported variously as Wallace Beery or Albert "Cubby" Broccoli, the future James Bond producer.) The postmortem examination concluded, however, that foul play was not a factor, and that Healy's death was due to toxic nephritis brought about by chronic alcoholism. See Cox and Terry, *One Fine Stooge*, 43–46; Jeff Lenburg, Joan Howard Maurer, and Greg Lenburg, *The Three Stooges Scrapbook* (Secaucus, NJ: Carol Publishing, 1999 [1982]), 13–17.

46. "Paramount, Newark," *Variety*, February 28, 1940, n.p.; "Colonial, Dayton," *Variety*, April 10, 1940, n.p., Three Stooges clippings file, BRTC.

47. Salary information taken from Cox and Terry, *One Fine Stooge*, 71.

48. William S. Holman (studio manager) to Jack Cohn, February 23, 1935, Columbia correspondence, 1935–1950, JWC. Production on *Three Little Pigskins* took five long days, totaling seventy hours work, whereas the Stooges' previous Columbia productions had been brought in within four. Production data taken from "1933–1934 Two Reel Program" and "1934–1935 Shorts Program," Columbia—Lists, ca. 1934–1958, JWC.

49. Bernds, *Mr. Bernds Goes to Hollywood*, 272.

50. "Jules's stuff was very violent," Ed Bernds explained. "That was just his way of doing it. His stuff was just violent. He used a lot things I didn't like and never used: the finger in the nose; ice tongs in the ears." Bernds quoted in Cox and Terry, *One Fine Stooge,* 47.

51. White quoted in Bruskin, *The White Brothers,* 243.

52. Howard quoted in Cox and Terry, *One Fine Stooge,* 41–42.

53. Tom Dardis, *Keaton: The Man Who Wouldn't Lie Down* (New York: Scribner's, 1979), 245; Marion Meade, *Buster Keaton: Cut to the Chase—a Biography* (New York: HarperCollins, 1995), 233; Okuda and Watz, *The Columbia Comedy Shorts,* 139.

54. Buster Keaton, with Charles Samuels, *My Wonderful World of Slapstick* (New York: Da Capo Press, 1982 [1960]), 253, 259.

55. The two films not directed by White were the series pilot, *Pest from the West* (June 1939), and *So You Won't Squawk* (February 1941; see n. 61 below), both directed by Del Lord.

56. Unidentified colleague quoted in Okuda and Watz, *The Columbia Comedy Shorts,* 25.

57. Schmidt, "Interview with Jules White," 44.

58. Meade, *Buster Keaton,* 235.

59. The allowance for Keaton's pictures was reduced for the 1940–1941 season to fifteen thousand. Production costs calculated from budgets in the folder Columbia—lists (costs), JWC.

60. Keaton, *My Wonderful World of Slapstick,* 253.

61. The only one of his later Columbia shorts in which Keaton did *not* appear with Ames was *So You Won't Squawk.* A short comedy remake of the Joe E. Brown Columbia feature *So You Won't Talk* (released the previous year), the film does not appear to have been written with Keaton in mind, which perhaps accounts for Ames's absence. (Unlike other scenarios in the series, the Keaton role is at no point referred to in the script as "Buster" or "Keaton"—instead, "Eddie.") "So You Won't Squawk," Final Continuity, October 18, 1940, JWC.

62. "What the Picture Did for Me," *MPH,* May 25, 1940, 60.

63. "The Taming of the Snood," final continuity, February 8, 1940, JWC.

64. "What the Picture Did for Me," *MPH,* July 13, 1940, 37.

65. See, for instance, David Macleod, *The Sound of Buster Keaton* (London: Buster Books, 1995), 75. For a perceptive reading of Keaton's films with Durante, see Joanna E. Rapf, "Mesh, Match or Blend: Buster Keaton's Films with Jimmy Durante," *New Review of Film and Television Studies* 5, no. 3 (December 2007): 317–31.

66. Noël Carroll, *Comedy Incarnate: Buster Keaton, Physical Humor, and Bodily Coping* (Malden, MA: Blackwell, 2007).

67. This articulation of the narrative/slapstick distinction as a contrast between Keaton's two female costars—the conventionally pretty Appleby versus the rough and brassy Ames—can also be found in *Nothing but Pleasure* and *The Taming of the Snood.* It relates to the long-established tradition in American comedy that permitted women to be laughable only to the extent that they deviated from classical standards of beauty. See Henry Jenkins, *What Made Pistachio Nuts? Early Sound Comedy and the Vaudeville Aesthetic* (New York: Columbia University Press, 1992), ch. 9; Kathleen Rowe, *The Unruly Woman: Gender and the Genres of Laughter* (Austin: University of Texas Press, 1995).

68. "General Nuisance," final continuity, March 5, 1941, JWC.

69. Carroll, *Comedy Incarnate,* 5, 10 (emphasis added); Tom Gunning, "Crazy Machines in the Garden of Forking Paths: Mischief Gags and the Origins of American Film Comedy," in Kristine Brunovska Karnick and Henry Jenkins, eds., *Classical*

Hollywood Comedy (New York: Routledge, 1995), 87–105. See also Tom Gunning, "Buster Keaton, or the Work of Comedy in the Age of Mechanical Reproduction," *Cineaste* 21, no. 3 (1995): 14–16; and Michael North, *Machine-Age Comedy* (New York: Oxford University Press, 2009), ch. 1.

70. Lewis Mumford, *Technics and Civilization* (New York: Harcourt and Brace, 1934), 10.

71. Gunning discusses these kinds of objects in his essay "Crazy Machines in the Garden of Forking Paths," where he sees them as precursors for the mechanical devices of later silent comedy. "Such devices," he writes, "provide the central heritage that early film comedy hands on to later silent comedy, from Lumière's garden hose . . . to Chaplin's assembly lines and Keaton's locomotives." (Gunning, "Crazy Machines in the Garden of Forking Paths," 98.) I am arguing, by contrast, for a distinction between the *tools* of early comedy and the *machines* of later slapstick, a distinction that, following Mumford, I locate in the object's degree of automatism vis-à-vis its human operator. See also Tom Gunning, "Mechanisms of Laughter: The Devices of Slapstick," in Tom Paulus and Rob King, eds., *Slapstick Comedy* (New York: Routledge, 2010), 137–51, esp. 150n10.

72. Bourdieu, *The Rules of Art*, 156 (italics in original).

73. "Movie of the Week: Hollywood Cavalcade," 63.

74. Review of *Sue My Lawyer*, MPH, n.d., n.p., "Columbia—Clippings," 8, JWC; review of *Pest from the West*, MPH, n.d., n.p., "Columbia—Clippings," 11, JWC.

75. Barbara Myerhoff, "Life History among the Elderly: Performance, Visibility, and Re-membering," in M. Kaminsky, ed., *Remembered Lives: The Work of Ritual, Storytelling, and Growing Older* (Ann Arbor: University of Michigan Press, 1992), 240.

76. Gene Fowler, *Father Goose: The Story of Mack Sennett* (New York: Covici-Friede Publishers, 1934), 180–81.

77. Maxine Block, "Hollywood Cavalcade," *Photoplay Studies* 5, no. 18 (1939): n.p., *Hollywood Cavalcade* clippings file, AMPAS. Keaton's only work with the former Keystone impresario in fact came in the sound era, in the Educational short *The Timid Young Man* (October 1935), on which Sennett worked as director for hire, having abandoned his studio operations to bankruptcy two years earlier.

78. Exhibitor reports from "What the Picture Did for Me," *MPH*, December 21, 1940, 53; May 25, 1940, 60; and April 4, 1942, 55 (emphasis added in all instances).

79. "What the Picture Did for Me," *MPH*, November 18, 1939, 65,

80. "What the Picture Did for Me," *MPH*, April 18, 1936, 72; and February 26, 1938, 60. In a similar vein, Stooges director Ed Bernds later recalled how the team's shorts were previewed only "in relatively unsophisticated places" like Glendale, still a relatively rural suburb of Los Angeles. Bernds, *Mr. Bernds Goes to Hollywood*, 296.

81. Quotes taken from "What the Picture Did for Me" clippings, *MPH*, n.d., n.s., "Columbia—Clippings," 4, 16, JWC.

82. Doherty quotes Universal producer Stanley Bergman in 1931: "We cannot make comedies that the children frown upon while the grown ups laugh. . . . That has been one of the great evils of the talking picture, and we must see that there are no more long-faced kids curled up in theatre chairs while some sophisticated two-reeler or other is being projected in front of their fathers and mothers." In Thomas Doherty, "This Is Where We Came In: The Audible Screen and the Voluble Audience of Early Sound Cinema," in Melvyn Stokes and Richard Maltby, eds., *American Movie Audiences: From the Turn of the Century to the Early Sound Era* (London: BFI, 1999), 152–53.

83. "What the Picture Did for Me," *MPH*, January 29, 1938, 79.

84. Gem Theatre manager quoted in Okuda and Watz, *The Columbia Comedy Shorts*, 11.

85. On ethnic and tramp stereotyping in vaudeville and early film, see my book *The Fun Factory: The Keystone Film Company and the Emergence of Mass Culture* (Berkeley: University of California Press, 2009), ch. 2.

86. New Deal–era redistributive economics also came in for ribbing in the Andy Clyde Columbia short, *Share the Wealth* (March 1936): a small-town mayoral candidate (Clyde) runs for office on a "share the wealth" platform only to change his mind and flee town when he inherits fifty thousand dollars.

87. Jules White's own copy of the *Horses' Collars* script has "Stooge #3" handwritten on the title page, suggesting that the film may in fact have been intended for slightly earlier production. Corrected script (29 pp.), *Horses' Collars*, undated, JWC.

88. Bland Johaneson, "Three Stooges' Expo: They'll Show You How at the Paramount," *New York Daily Mirror*, September 17, 1938, n.p., Three Stooges clippings file, BRTC.

89. Howard McClay, "These Stooges Are No Dummies," *Los Angeles Daily News*, August 6, 1953, n.p., Three Stooges clippings file, AMPAS.

90. The concept of ideologeme is most fully developed in Jameson's classic text, *The Political Unconscious: Narrative as a Socially Symbolic Act* (Ithaca, NY: Cornell University Press, 1981), where it is defined as a "historically determinate conceptual or semic complex which can project itself variously in the form of a 'value system' or 'philosophical concept,' or in the form of a protonarrative, a private or collective narrative fantasy" (115).

91. On the "bomb under the bed" and "ghost in the pawnshop" plots, see Hilde D'haeyere, *Dislexicon of Slapstick Humor, Funny Cinematography, and Very Special Effects* (Ghent, Belgium: MER. Paper Kunsthalle, 2011), 8, 25. On the recycling of gags in film slapstick, see Bryony Dixon, "The Good Thieves: On the Origins of Situation Comedy in the British Music Hall," in Paulus and King, eds., *Slapstick Comedy*, 21–36; and Anthony Balducci, *The Funny Parts: A History of Film Comedy Routines and Gags* (Jefferson, NC: McFarland, 2012). It is worth adding that Columbia's veteran filmmakers often took this recycling to shamelessly self-plagiarizing extremes, mining their own back catalogs not just for basic comic situations but for entire plots. Del Lord, for example, utilized the identical narrative of a Taxi Boys short he had directed for Roach, *Wreckety Wrecks* (February 1933), for the El Brendel short *Ready, Willing, but Unable* (May 1941). Clyde Bruckman's script for the Andy Clyde short *Andy Plays Hookey* (December 1946) was simply a two-reel condensation of *The Man on the Flying Trapeze* (1935), a W.C. Fields feature directed by Bruckman. At times, the unit's top brass seemingly advised caution, as when a proposed race-to-the-rescue climax for a Stooges' short was jettisoned after Jules White noticed that "H. Lloyd did most of this" (in the 1924 feature *Girl Shy*). At others, more caution would still have been wise: Lloyd in fact eventually sued Columbia after Bruckman copied scenes from Lloyd's *Movie Crazy* (1932)—which, again, Bruckman had directed—for the Three Stooges short *Loco Boy Makes Good* (January 1942). The annotation "H. Lloyd did most of this" is in "Oily to Bed, Oily to Rise," unused script for chase scene, n.d., JWC. On Lloyd's suit against Columbia, see Okuda and Watz, *the Columbia Comedy Shorts*, 32–33.

92. John Crosby, "Return of The Stooges," *NYHT*, February 1, 1959, 9–10, Three Stooges clippings file, BRTC.

93. For a discussion of dime-novel fiction in these terms, see Michael Denning, *Mechanic Accents: Dime Novels and Working-Class Culture in America* (London: Verso, 1987), esp. chs. 6 and 8; on their application to early film slapstick, see my book *The Fun Factory*, 96–104.

94. Denning, *Mechanic Accents*, 81.

95. Warren I. Susman, "The People's Fair: Cultural Contradictions of a Consumer Society," in *Culture as History: The Transformation of American Society in the Twentieth Century* (New York: Pantheon Books, 1973), 212.

96. Burke quoted in Susman, "The People's Fair," 211.

97. Susman, "The People's Fair," 212.

98. Michael Denning, *The Cultural Front: The Laboring of American Culture in the Twentieth Century* (London: Verso, 1997), 124–25.

99. On Temple, see, in particular, Charles Eckert's famous essay, "Shirley Temple and the House of Rockefeller," *Jump Cut* 2 (1974): 1, 17–20. On Will Rogers, see Lary May, *The Big Tomorrow: Hollywood and the Politics of the American Way* (Chicago: University of Chicago Press, 2002), ch. 1.

100. "Laurel and Hardy Reach Pinnacle of Success" and "Laurel and Hardy Success Due to Co-operation," in *Pardon Us* pressbook, 4, BRTC.

101. "'Our Relations' Marks 10th Year Together for Laurel and Hardy Team," *Our Relations* pressbook, 5, BRTC.

102. "Laurel and Hardy Success Due to Co-operation," 4.

103. B. R. Crisler, "Random Notes on Pictures Persons," *NYT*, June 26, 1938, n.p., Three Stooges clippings file, BRTC.

104. "Hoi Polloi," script, n.d., 29–30, JWC.

105. "Cash and Carry," fifth draft script, April 28, 1937, 2, JWC.

106. Ibid., 8–9.

107. Ibid., 32–33. This is not quite how the actual film ends. Curly's line "Gee, Mr. President, you're a swell guy" is moved to the end after Moe violently corrects his misunderstanding of clemency. Moe then adds, "You said it," and the film fades out with the Stooges saluting the commander-in-chief.

108. This failure of resolution is nowhere more pronounced than in *Loco Boy Makes Good*, in which the boys put on a nightclub show to raise money to cover the widow's mortgage. The film's nightclub scenes degenerate into a chaos that utterly fails to complete the charity motif, instead concluding abruptly with Curly whirling an escaped skunk around his head, the actual plot seemingly forgotten.

109. Henri Bergson, *Laughter: An Essay on the Meaning of the Comic*, trans. Cloudesley Brereton and Fred Rothwell (London: Macmillan, 1911); Mikhail Bakhtin, *Rabelais and His World*, trans. Helen Iswolsky (Cambridge, MA: MIT Press, 1968).

110. Lowell E. Redelings, "The Hollywood Scene," *Hollywood Citizen-News*, June 9, 1952, n.p., Jules White clippings file, AMPAS.

111. White quoted in Bruskin, *The White Brothers*, 265.

112. Production data derived from "Lists—Costs," JWC.

CODA

1. Program, *Hollywood Cavalcade* clippings file, AMPAS.

2. "The expression 'prat falls' [sic] is not acceptable, and must be changed." Joseph Breen to Jason S. Joy, May 15, 1939, *Hollywood Cavalcade*, PCA Files, AMPAS.

3. Which is to say that the films' historiographic mode is that of *comedy*, as one of the four modes of emplotment in traditional historical writing identified by Hayden White (the others being romance, tragedy, and satire). See White, *Metahistory: The Historical Imagination in Nineteenth-Century Europe* (Baltimore, MD: John Hopkins University Press, 1973).

4. "Movie of the Week: Hollywood Cavalcade," *Life*, October 9, 1939, 63–69.

5. Press releases, June 13, 1939 and June 27, 1939, *Hollywood Cavalcade* clippings file, AMPAS.

6. See ch. 5.

7. Barbara Myerhoff, "Life History among the Elderly: Performance, Visibility, and Re-membering," in M. Kaminsky, ed., *Remembered Lives: The Work of Ritual, Storytelling, and Growing Older* (Ann Arbor: University of Michigan Press, 1992), 240. See ch. 5 for the further applicability of this concept to slapstick nostalgia.

8. Review of *Hollywood Cavalcade*, *Variety*, October 4, 1939, n.p., *Hollywood Cavalcade* clippings file, AMPAS.

9. Review of *Hollywood Cavalcade*, *Film Daily*, October 4, 1938, n.p. *Hollywood Cavalcade* clippings file, AMPAS.

10. Terry Ramsaye, review of *Hollywood Cavalcade*, *MPH*, October 7, 1939, 35.

11. The concept of "interpreted" nostalgia is suggested by Fred Davis, who proposes three "orders" of nostalgia: first-order or simple nostalgia ("a positively toned evocation of a lived past"); second-order or reflexive nostalgia (characterized by "empirically oriented questions concerning the truth, accuracy, completeness, or representativeness of the nostalgic claim. Was it really that way?"); and third-order or interpreted nostalgia (characterized by "analytically oriented questions concerning its sources, typical character, significance, and psychological purpose. Why am I feeling nostalgic?"). Fred Davis, *Yearning for Yesterday: A Sociology of Nostalgia* (New York: Free Press, 1979), 18, 21, 24.

12. Review of *Hollywood Cavalcade*, *Variety*, October 3, 1939, n.p., *Hollywood Cavalcade* clippings file, AMPAS.

13. "The Hollywood Scene," *MPH*, October 21, 1939, 32.

14. "Keystone Comedy Revived," *MPH*, October 28, 1939, 26.

15. "Need Long Laughs, Says Max Gordon," *MPH*, October 21, 1939, 54.

16. "Trend toward Revival of Old Slapsticks," *MPH*, June 1, 1940, 51.

17. David Kalat, "The History of the History of Silent Comedy," *Movie Morlocks*, February 18, 2012 http://moviemorlocks.com/2012/02/18/the-history-of-the-history-of-silent-comedy (accessed July 2, 2015).

18. This point, and the subsequent examples, are from Henry Jenkins and Kristine Brunovska Karnick, "Introduction: Golden Eras and Blind Spots—Genre, History and Comedy," in *Classical Hollywood Comedy* (New York: Routledge, 1995), 1–2.

19. James Agee, *A Death in the Family* (New York: Grosset and Dunlap, 1967 [1957]), 14.

20. Walter Kerr, *The Silent Clowns* (New York: Da Capo Press, 1980 [1975]), 10.

21. "SG on Slapstick Kick in Rebirth Of '3 Stooges,'" *Variety*, March 21, 1959, n.p., Three Stooges clippings file, BRTC.

22. "'3 Stooges' Set Pace at WGN-TV," *Variety*, January 29, 1959, n.p., and "Three Stooges, Omaha, Aug. 24," *Variety*, August 30, 1961, n.p., Three Stooges clippings file, BRTC.

23. "4 Clowns," *MPH*, June 17, 1970, n.p., *4 Clowns* clippings file, AMPAS. Youngson's *When Comedy Was King* cost around $100,000 to produce and, within six months, drew some $218,000 in domestic rentals alone. "Youngson Anthologies of Silents Continue Showing Coin Potential," *Variety*, October 14, 1960, n.p., Robert Youngson clippings file, AMPAS.

24. Bob Salmaggi, "The Three Stooges: What Price Violence?," *NYHT*, July 10, 1960, 59.

25. Salmaggi, "The Three Stooges," 58.

26. Susan Stewart, *On Longing: Narratives of the Miniature, the Gigantic, the Souvenir, the Collection* (Durham, NC: Duke University Press, 1993), 66 (emphasis in original).

27. According to Sergei Eisenstein, the potent effect of cartoons stemmed from their quality of "plasmaticness"; that is, the way in which cartoon figures behave "like the primal protoplasm, not yet possessing a 'stable' form, but capable of assuming any form." Sergei Eisenstein, *Eisenstein on Disney*, trans. Jay Leyda (London: Methuen 1985), 21. On miniaturization, see my earlier discussion of *Babes in Toyland* (1934), ch. 4.

28. See the Blackhawk Films Collection, AMPAS. Of the approximately hundred letters in the collection—on topics ranging from queries about silent versus sound speed, print care, and the like—twenty-five are substantively concerned with the company's live-action comedy releases.

29. Kalton C. Lahue and Sam Gill, *Clown Princes and Court Jesters: Some Great Comics of the Silent Screen* (South Brunswick, NJ: A.S. Barnes, 1970); Kalton C. Lahue with Terry Brewer, *Kops and Custards: The Legend of Keystone Films* (South Brunswick, NJ: A.S. Barnes, 1968); Kalton C. Lahue, *Mack Sennett's Keystone: The Man, the Myth and the Comedies* (South Brunswick, NJ: A.S. Barnes, 1971), and *Collecting Classic Films* (New York: American Photographic Book Publishing, 1970). Lahue's remarkable hot streak of publications from the 1960s and early 1970s also includes *Continued Next Week: A History of the Moving Picture Serial* (Norman: University of Oklahoma Press, 1964), *Bound and Gagged: The Story of the Silent Serials* (South Brunswick, NJ: A.S. Barnes, 1968), *Dreams for Sale: The Rise and Fall of the Triangle Film Corporation* (South Brunswick, NJ: A.S. Barnes, 1971), *Ladies in Distress* (South Brunswick, NJ: A.S. Barnes, 1971), *Winners of the West: The Sagebrush Heroes of the Silent Screen* (South Brunswick, NJ: A.S. Barnes, 1971), and *World of Laughter: The Motion Picture Comedy Short, 1910–1930* (Norman: University of Oklahoma Press, 1972). He subsequently went on to write automobile manuals, no less prolifically.

30. Stewart, *On Longing*, 151, 138.

31. Ibid., 151 (emphasis in original).

32. See, for instance, Amelie Hastie, *Cupboards of Curiosity: Women, Recollection, and Film History* (Durham, NC: Duke University Press, 2007), 24–28.

33. Brent E. Walker, *Mack Sennett's Fun Factory: A History and Filmography of His Studio and His Keystone and Mack Sennett Comedies, with Biographies of Players and Personnel* (Jefferson, NC: McFarland, 2010); Richard Roberts, with Robert Farr and Joe Moore, *Smileage Guaranteed: Past Humor, Present Laughter—Musings on the Comedy Film Industry 1910–1945*, vol. 1, *Hal Roach* (Phoenix, AZ: Practical Press, 2013). Steve Massa's work is represented both in single-author publications and in his and Ben Model's coproduced DVDs and recurrent "Cruel and Unusual Comedy" series at New York's Museum of Modern Art. Massa's publications

include *Lame Brains and Lunatics: The Good, the Bad, and the Forgotten of Silent Comedy* (Albany, GA: Bear Manor Media, 2013), *The Mishaps of Musty Suffer: DVD Companion Guide* (New York: Undercrank Productions, 2014), and *Marcel Perez: The International Mirth-Maker* (New York: Undercrank Productions, 2015). Massa and Model's DVDs include three volumes of *Accidentally Preserved* (New York: Undercrank Productions, 2013, 2014, 2015), two volumes of *The Mishaps of Musty Suffer* (New York: Undercrank Productions, 2014, 2015), *The Marcel Perez Collection* (New York: Undercrank Productions, 2015), and *Found at Mostly Lost* (New York: Undercrank Productions, 2016), among others.

INDEX

CPSIA information can be obtained
at www.ICGtesting.com
Printed in the USA
LVOW03s0705070917
547806LV00003B/5/P